A New Farm Language

How a sharecropper's son discovered
a world of talking plants, smart insects,
and natural solutions

By W. Joe Lewis

A New Farm Language

How a sharecropper's son discovered
a world of talking plants, smart insects,
and natural solutions

By W. Joe Lewis

Acres U.S.A.
Greeley, Colorado

A NEW FARM LANGUAGE

Acres U.S.A.
P.O. Box 1690
Greeley, CO 80632
970-392-4464
info@acresusa.com • www.acresusa.com

Printed in the United States of America

Book Editors: Ben Trollinger and Ryan Slabaugh
Illustrations: Vincent A. Keesee, Ph.D
Book Designer: Carl Chiocca

Publisher's Cataloging-in-Publication Data
Names: Lewis, Joe, author.

Title: A new farm language : how a sharecropper's son discovered a world of talking plants, smart insects, and natural solutions

Description: Agricultural and biographical | Greeley, CO: Acres U.S.A., 2021

Identifies: ISBN print 978-1-60173-138-8; eBook 978-1-601730139-5

Subject: LCSH Agriculture—Biography | Environmental Sciences | Nature—Social aspects | American South | South, The | Cotton—Environmental Aspects | Cotton—Disease and pest resistance

Classification: S604.5-604.64 | DDC 638

ISBN-10: 1-60173-166-3

ISBN-13: 9781601731661

Dedication

In memory of my parents, Ferd and Grace Lewis:
Your life close to the land, unencumbered by the lure of materialism, and your deep love of God, family, and neighbors, have taught us all the profound power and lasting impact of a simple Godly life.

To my wife, Beth:
Your abiding companionship, zest for life, boundless energy, and caring outreach to others, inspire and give me energy every day.

To my children and grandchildren (Alan, Joanne, and their families), and in memory of Jay:
No father or grandfather could ask for more. From our laughter together with your first bounces on my knee to your loving respect during adulthood, each of you has been an endless source of joy and happiness, as we have celebrated in times of merriment and rallied together in times of need.

In memory of my friend and colleague
Vincent Keesee (1935-2021):
Special thanks to the family for the permission to use his illustrations in this book.

Contents

Acknowledgements

My friend, Vincent A. Keesee, Ph.D., a retired professor of art of a local college graciously contributed the sketches included throughout the book. This contribution started with an inquiry from me as to whether he would prepare a couple of sketches that I had in mind for the first couple of chapters. However, when I provided him a draft copy of the book he took a personal interest and provided the contributing material included throughout the entire book. His wife, Marianna, in the meantime read the book draft and provided me very encouraging feedback and suggestions. I am especially thankful. Dr. Keesee is well-known, gifted artist, and his work is highly coveted by many. He recently released a book himself entitled, *Hallelujah: A Lifetime of paintings.*

The narrative and perspectives herein represent relationships of many people who have contributed extensively to the enrichment of my life. Many of those people, and their roles, are described within the pages of this book, hopefully in a manner that reflects my appreciation and respect.

Moreover, I am deeply appreciative for the support and assistance of numerous people in moving the book from an initial idea to reality:

My wife, Beth, provided untiring encouragement and

assistance in all aspects of the project. She shared ideas and suggestions during initial planning and formulation of the book's scope and central topic; she patiently helped ensure that I had time and space for all aspects of the work; she provided feedback on drafts; and, perhaps most importantly, she steadily helped me maintain it as a priority on my schedule. I also owe special thanks to my children and grandchildren for their encouragement and support. They provided much of the motivation for writing the book in the first place.

My sister, Patsy Kelly, and cousin, Charolette Wallace, helped immensely in sharing memories, and in reading and providing suggestions, especially for the early chapters.

Harold Chambers, a friend and colleague who worked closely with Beth and me on a variety of educational/community projects came to know about my background a number of years ago. Through that information and related discussions, he strongly urged me to write a book on the subject, and was perhaps the first to spark a serious contemplation of the undertaking.

Numerous friends, including Allen and Sherron Mull, Scott and Bunny Chestnutwood, Bob Cartwright, and Johnny Johnson, provided valuable encouragement through their ongoing interest in the book's progress, reading sample parts and providing feedback and cheering me on.

In addition to the cooperative work covered in the book, several of my professional colleagues reviewed the book and offered valuable feedback. These colleagues include Jim Tumlinson, Alton Walker, Dawn Olson, Glen Rains, Felix Wackers, and Richard Jones.

Within the book, I name many of the valuable professional cooperators I enjoyed throughout my years of worked with the USDA-ARS at the Tifton location. Our programs were blessed with a continuous array of excellence in management, administrative, and technical staff. I cannot name them all, but those that come to mind include: Alton Sparks, Charlie Rogers, Bob Lynch, Lillie Brady, Kay Carr, Lynn Henderson, Deborah Osborne, Roger McClain, Debbie Padgett, Tom Maze, Bob Williams, David Webb, Larry Walters, Simmy McKeown, Thoris Green, Lloyd Copeland, Ricky Fletcher, Harlene Kegley, Amy Heidt, Debbie Waters, Faye Childers, Zoe Chung, Margret Woods, Joanne Denham, Marcie Lathan, and Tommy Hester. In summary, USDA-ARS was just an excellent organization to work for at all levels, national to local.

Also, the cooperation and support that I received from the University of Georgia, both from the professional and support staff were superb. In addition to those named specifically in the text, I enjoyed scientific collaboration with Bob Matthews, Jim Todd, Gary Herzog, and Paul Martin. Since considerable amounts of funding, programmatic, and personnel support came through the University of Georgia, considerable assistance was required in the way of personnel actions, record keeping, and budgetary activities. The bulk of this assistance was channeled through the Entomology Department at the Tifton Campus with full cooperative support of all the Department Heads over the years, starting with Dr. T. Don Canerday and followed by Drs. Max Bass, Gerald Musick and others. These needs were directly handled always in a courteous and efficient manner largely by Elaine Belk, Carol Ireland, and others, including the support of staff in the international Programs Division of the University of Georgia. Also, the full support of the Station Director or Assistant Dean, depending on the organizational structure, was also needed and was steadily provided over the years by Drs. Frank King, Broadus Brown, W. C. McCormick, Gale Buchanan, Philip Utley, David Bridges, and Joe West.

Over the most of my career I also held an Adjunct/Courtesy Professor appointment with the University of Florida. I enjoyed similar excellent support and cooperation with that institution. It was particularly of value in my cooperation with Jim Tumlinson in our work together.

For several years I functioned as a satellite site of the USDA-ARS Gainesville location, during which time I received excellent management support from Herb Oberlander and Carrol Calkins.

Foreword

It was the spring of 2013 when Joe Lewis changed my mind.

I was researching my second book, *The Dorito Effect*. Which is to say I was in a state of curiosity-fueled confusion, fumbling my way down a path of inquiry that seemed to have no discernible destination.

Too much science will do that to you. I had, for years, been immersing myself in research based on crisp methods and observations. The myths and biases of my former self had all been obliterated by evidence-based reasoning and statistical significance.

And it was becoming a problem. I was studying flavor. I was trying to understand why food tastes the way it does, why people like it, or do not like it. You could say I was studying life itself, because there is no act more inherently biological and organic than that of eating.

And yet, all this life I was studying had come to seem rather lifeless. Humans, I was told time and again, are programmed to pursue calories. That's all eating was—a never-ending hunt for energy. The rich, dazzling tastes, flavors and textures we experience when we eat were, apparently, some grand illusion, beneath which lies a crisp little game of calories. It was elegant and compellingly simple, as good science should be. And yet, it made me feel cold and numb.

"I would be pleased to speak with you," is how Joe responded when he received an email out of the blue from a journalist in Toronto. I phoned him that very day at his house in Tifton, Georgia, and he spent the better part of an afternoon telling me a story about science that wasn't so simple. Joe described how an organism as regular as a cotton plant is as strategic as the most cunning general, that when it is attacked by a caterpillar, it sends out a chemical SOS. It warns its fellow cotton plants of the imminent danger. Even more amazingly, it calls in the Air Force—little wasps that lay their eggs in the offending caterpillars, eggs that hatch and consume their hosts from the inside out.

In the space of a few hours, a farmer's field went from a picture of bucolic serenity to a place of ongoing war, with attacks and defenses, enemies and allies, payoffs and death. And that war was fought with chemicals. A compound that meant "danger" to a plant meant "food" to a flying insect. And these wasps weren't little organic robots, drawn by birth to this or that compound. They learned. They saw about them a world of smell, colors and shapes.

Joe taught me a new way of seeing the world. Flavor, I would come to understand, wasn't simply the mere taste of food. It is the chemical language by which nature conducts its business. When we taste food, any food—a grape, a chicken wing or a leaf of mint—we taste the theater of life itself.

In the pages of this remarkable and compelling book, Joe shares his journey, one that began in nature, in the backwoods of Mississippi, and ultimately is about nature. In a voice that is innocent and curious, yet also wise, he shows how the natural world works in ways that defy our often simple ways of thinking. Eons of evolution have found solutions to problems we have only recently stumbled into.

Joe Lewis taught the world about how plants talk to insects. But he gave us something much more important: the wisdom to see the wonder in creation. Too often, we think of that wonder as the enemy of science—that it must be demystified, controlled and reduced into numbers. Joe shows us that just isn't true. Wonder is not the adversary of science. It is, and will always be, its fuel.

—Mark Schatzker, author
www.MarkSchatzker.com

Introduction

I was surprised, and pleased, when I received a phone call from Joe Lewis asking if I would be willing to read and review his manuscript for a new book. I remembered him from when I was on faculty of the University of Georgia in the late 1980s. I remembered he was working on some interesting new research at the USDA research station at Tifton, Georgia about the ability of plants to protect themselves from insects. I also remembered a presentation he had made about his work in community development with the city of Tifton. I didn't recall ever having a conversation with him and was surprised that he knew enough about me to know I would be interested in reading his book. After I read Joe's manuscript, I was even more flattered that someone with his distinguished record of scientific accomplishments had asked me to review his book. We have had many phone conversations since. So, I don't think he will mind my referring to him simply as Joe rather than Dr. Lewis.

As I began reading, I was struck the similarity in our life stories: our early years on the verge of poverty without electricity or indoor plumbing, our hard work in the fields of farms that could barely support a family, but also our opportunities to attend state universities and to pursue careers in agricultural research, education, and extension. Joe seemed to know from the beginning that he wanted to learn more about the relationships between the plants and insects he had observed in the cotton fields as a child. It took me longer to finally find my academic niche in sustainable agriculture. Joe's research eventually brought him national and international acclaim from the scientific community.

After I finished reading the book, I could understand why Joe had asked me to review it. Our academic disciplines were very different, but our conclusions regarding how the world works and our place as humans within it are much the same. He was a "hard scientist" conducting controlled experiments. I was a "social scientist" doing applied research and extension work. Yet our ways of thinking about the problems and opportunities in agriculture, science, community, society, and life are obviously very much in harmony. We both have come to the conclusion that were we are working and living in an integrally interconnected world in which everything we do affects everything else—some in small ways, in others, large. We can't do just one thing.

Our world works like a big, complex living organism, and we humans are inseparable parts of the larger organismic whole of the earth. It's impossible even to separate science from the scientists. That's why Joe begins his book by describing his observation as the son of a sharecropping family in the cotton fields of Mississippi. The relationships between plants and insects he observed in the fields as a child informed his work in the laboratory as a scientist. That's why he writes about his memories of conversations with other family members and among neighbors at the country store and in the communities. At the most basic level, the relationships among people in families, communities, and in society, are much like the relationships among the multitudes of living plants, insects, and other organisms in the fields. Plants, insects, families, communities, scientists; we are all parts of inseparable wholes, and ultimately the whole of the earth. Whatever affects one, affects the whole.

If this were an ordinary book about important scientific discoveries, it would have focused on the process of discovery, verification, and implications for application and commercialization of results. All of these elements are in the book, but its scope is much broader and its implications much deeper. An ordinary book would have focused more narrowly on the work of the author. Joe writes about his relationships with those who helped him develop as a scientist. He generously shares credit for his discoveries with a whole team of scientists who knew each other personally and worked collaboratively to advance scientific understanding. If this were an ordinary book, a major section would have been devoted to how smart wasps could be used as organic or biological inputs in commercial farming operations. Instead, Joe writes about developing a "Year-round Ecosystem Management System" in which nature provides smart wasps, talking plants, and a host of other benefits for free. In Joe's system, chemical pesticides are sometimes justified, but only as a temporary intervention until a healthy, balanced agroecosystem can be restored.

Joe doesn't shy away for using the terms sustainable and non-sustainable in describing his total systems approach to pest management. He writes that sustainable farming systems rely on the internal strengths of natural ecosystems, which allow them to minimize external interventions. Non-sustainable farming systems rely heavily on external interventions, which eventually destroy the internal strengths of agroecosystems. The industrialization of agriculture has destroyed critical relationships among living organisms in soils, among insects, plants, and animals on farms, which has diminished the internal strength of agroecosystems and now threatens the sustainability of American agriculture. Throughout the book, Joe makes references to how the transition from diverse family farms to large-scale industrial farming operations has diminished the sustainability of agriculture and degraded the quality of life and rural communities.

In the latter chapters of the book, Joe returns to his personal reflections on the importance of relationships in his own family and community. He explains how a loss of appreciation for the importance of social relationships in rural communities has paralleled the loss of appreciation for the importance of ecological relationships in agriculture. He writes of his personal experiences in Tifton, Georgia, and how his work toward sustainable agriculture in the laboratory has paralleled his work toward localizing purpose-based education and sustainable development in his own community. He writes about how relationships in his own family have helped him understand the importance of sustainable families to sustainable communities and societies. Joe concludes his book with a message of hope. He explains how new technological discoveries could focus on expanding the built-in strength of nature, rather than expanding external interventions. He explains how these same principles and possibilities also apply to strengthening families, communities, and societies.

In the final chapter Joe describes a return visit to his homeplace in rural Mississippi and reflects on his life there and the consequences of a lost sense of connectedness among the people and between people and nature. He realizes that he might not have understood or appreciated the importance of relationships between wasps and plants in the laboratory if he had not lived with those relationships as a boy in those fields. He might not have understood the parallel between the negative impacts of interventionist industrial agriculture and interventionist rural economic development if he had not grown up in a socially vibrant rural community. Perhaps this is the reason Joe and I have reached similar conclusions regarding how the world works and where we fit within it. We both grew up in nature and were integrally interconnected with nature in our day-today lives. We knew intuitively that we were a part of nature and thus must have some purpose within that larger whole nature. Scientists are inseparable from their science—regardless of claims to the contrary. The separation of scientists from nature has led to science separated from reality.

Joe concludes his book by writing of the ability of people to rediscover a true sense of purpose and meaning in life by reconnecting with each other and with the earth. We are all part of the same interconnected web of life and find meaning in our lives by contributing in some way to the betterment of the web of life as a whole. He concludes, "This will require, above all else, a reconnection with nature. If I am saying nothing else here, I am saying that our disconnect from nature—her beauty, her power, her amazing ability to give—is, more than anything, the greatest threat to our survival." Joe Lewis' book, *A New Farm Language*, is not an ordinary book about a scientific discovery; it is a unique and important book about nature and the nature of life.

—John Ikerd,
  Professor Emeritus of Agricultural & Applied Economics
  University of Missouri Columbia
  College of Agriculture, Food and Natural Resources

"In nature nothing exists alone."

— Rachel Carson, *Silent Spring*

Preface

In the Name of Progress

The hum of jet engines, mixed with the aroma of freshly brewed coffee, greeted my first moment of awareness. "Wake up, baby," came my wife's voice as she nudged me from her window-side seat. "They're serving breakfast, and you missed the movie." It was May 30, 2008, and we were near our initial approach to Atlanta. Beth

Beth and I return from Israel in 2008, where I received Wolf Prize in Agriculture.

and I were returning from a week in Israel. What a week, indeed! I had been notified in January of my selection for the Wolf Prize, sponsored by the Wolf Foundation of Israel. This prize promotes achievements for the benefit of mankind in the sciences and arts. Modeled after the Nobel Prize, the award is presented annually in the fields of agriculture, chemistry, mathematics, medicine, physics, and the arts. The 2008 Wolf Prize recipients had gathered in Jerusalem that week for the official presentations. During an elegant ceremony at the Knesset, the president of the State of Israel, Shimon Peres, personally presented each laureate their certificate and monetary award.

President of Israel, Shimon Peres, presents me the Wolf Prize at a ceremony in Jerusalem.

My colleagues, Dr. James H. Tumlinson (Pennsylvania State University) and John A. Pickett (Rothamsted Research Center, UK), and I shared the award in the field of agriculture. Our certificate read: "For remarkable discoveries of mechanisms governing plant-insect and plant-plant interactions. Their scientific contributions on chemical ecology have fostered the development of integrated pest management and significantly advanced agricultural sustainability." The honor represented a major capstone to my forty-plus-year career as a research entomologist. It was gratifying to be returning home to Tifton, Georgia as a Wolf Laureate.

While sipping my morning coffee, I peered out of the airplane window at my native land underneath. The weather was clear, so the 30,000-foot view captured the immense natural grandeur of the eastern seaboard and Appalachian Mountains. America's strategic location between two great oceans, the presence of two vast mountain ranges, along with an enormous river network and basin gracing its central core, help provide a bountiful existence of forests, fertile farmlands, and diverse climatic conditions. What a blessed land of plenty and of opportunity, I thought, reflecting on how I had partaken liberally of these attributes. I am truly a product of the American Dream. I grew up in a very rural area of south Mississippi as a sharecropper's son grubbing out a living on borrowed land. For years, our entire material well-being, beyond the food that we grew directly, was dependent on a five-acre cotton crop worked by a single family mule, a couple of plows, and hand tools. My father could not read or write a single word. During much of that period, we had no electricity or running water and no form of automated transportation. Unencumbered by modern devices, our existence was centered on relationships. We loved generously and cared deeply. Life was simple. Life was rugged.

Somehow, I found my way into a very different domain. Through the benefits of free, public education, and a series of scholarships, I was able to obtain a strong education, culminating with a PhD in entomology, and embark on my career in scientific research, thereby participating in the front end of technological advances. That morning above Atlanta, looking out at the land below, I reflected on a key phrase in the Wolf Prize citation: "Their scientific contributions…have significantly advanced agricultural sustainability." That statement was especially meaningful. I had become concerned about the escalating issue of sustainability at both the national and global level.

So, I began to think about those issues and how they had become so pressing. A well-known ecological principle is that the universe is made up of systems within systems, whereby humankind and all other organisms are inextricably linked by two unifying forces: 1) the proper cycling of essential materials (i.e. oxygen, carbon, hydrogen, calcium, phosphorous, nitrogen, etc.); and 2) the capture and transfer of energy necessary to power the activity of living organisms, ultimately from the sun through photosynthesis by green plants.

All these systems have built-in mechanisms that enable them to maximize efficiency, self-regulate, self-renew, maintain balance, and minimize undesired variables. Examples include our immune system and our ability to regulate our body temperature. The best methods for fostering the well-being of any system are based on leveraging these intrinsic features. The built-in mechanisms should always be the first team for managing undesired occurrences in a system. External interventions or fixes are strictly backups that should be used sparingly and only for a short term. Dealing with a pest outbreak in an agricultural crop, for example, should start by asking, why is this pest a pest? What adjustments in the soil health, plant health, plant mixes, or landscape design should be made? Is there something interfering with the numbers and performance of natural enemies? Use of pesticides or similar external interventions should come only as a backup.

This guideline is universal and holds true for all other sorts of systems. In the case of human health, the first step should be potential corrections through nutrition, exercise, and/or rest before resorting to such treatments as antibiotics and pain killers. External interventions are not part of the system, so they don't operate within the system's feedback loops. Consequently they create disruptions and start becoming neutralized by the systems, thus requiring more to get the same results.

For generations we sensed these laws and generally worked within them, often by necessity because nature's methods were all that we had. Farmers were in many ways the world's first ecologists, though more as a matter of art than science. Our farming communities were made up primarily of small-scale family subsistence farms, and were true ecosystems of diverse multi-crop, livestock mixes. They were protected and passed along as family assets. Although their objectives have always been to manipulate these systems to maximize the harvests, farmers had vested interests in respecting and preserving the long-term health of their farms. Thus, they were careful to foster these components of the system by such practices as crop rotation, intercropping, recycling of residue, use of cover crops, and landscape ecology. Through folk wisdom passed down over generations, farmers knew this diversity promoted natural enemies and balance. Their inputs were often organic in nature, using elements like chicken litter for fertilizer because that was what was available. That was the context for our farming during my early years.

By the end of the Great Depression, however, near the time of my birth in 1942, key innovations and technology from the Industrial Revolution began to find their way into agriculture and all aspects of our lives (though arriving somewhat late for my family). With this revolution came healthcare, transportation, personal comfort, communication, automated machinery, and an ever-expanding array of chemical and electronic products. In the short term, these innovations provided much in the way of prosperity and an improved quality of life. But, as time moved along, the hidden price tag on these enhancements began to emerge. In agriculture, large monoculture farms using tractors, synthetic fertilizers, and pesticides replaced small, diverse, multi-crop farms. Pesticide resistance and contamination began to develop. Scientists began to report examples of species extinction. Urbanization started to escalate; the number of family farms decreased and small rural towns declined.

Science is often reductionist in nature and divides the study of systems into specialized parts. So, these technologies were developed and delivered in corresponding specialized ways. An item of machinery or chemical product, for instance, would often be developed and marketed for use on one crop, such as cotton, versus an overall cropping system. These approaches encouraged movements toward large-scale monoculture farming.

These practices and trends have continued to the point where traditional multifunctional family farms that historically functioned as an interactive part of the local community are now almost non-existent in the U.S. and much of the world. Rather, the organizational modes of centralization and specialization have brought about sprawling farms that, though they may be family-owned in many cases, are competing as businesses on a global basis and in specific commodity markets. At the same time, there is a disturbing pattern toward centralization of ownership and operation of all phases of the agricultural industry, including production, processing, marketing, and sometimes even the land itself. For any particular farm commodity, these various centers of production, processing, and marketing may be substantially separated geographically. On average, a meal now travels 1,500 miles from farm to table. Significant concern has emerged regarding how these trends, including the mergers of supermarkets, are affecting the security of our food supply and its quality, availability, distribution, and pricing. At the same time, centralization of sales and services to farms such as machinery, seeds, fertilizers, pesticides, and other supplies, coupled with urbanization, has brought about a widespread decline of the economic and social viability of most agriculture-based small towns. Most of these towns are now riddled with socioeconomic issues including poverty, low school graduation rates, lack of healthcare, and the disappearance of the qualified workforce that would be necessary to reverse these trends.

High input of therapeutic interventions dominates on-farm practices in direct opposition to the principles of self-regulation and self-renewal. Large-scale monoculture, together with clean tillage, lack of cover crops, and residue recycling has reduced biodiversity and soil health while increasing erosion and pollution of our soil and water. Escalating pesticide and fertilizer use, stemming from mounting resistance and ecosystem disruptions, has created the proverbial treadmill, where more and more of the same, or shifts to new products in an endless string of new selections, is required to get the same result. It seems that many of our best-trained agricultural leaders continue to see the problem as a weakness of the tool being used rather than finding and correcting any weakness in the system. This "game without end" of products bombarding the systems supports an ever-growing global agricultural chemicals market, where the solution has become the problem. The inherent strength of our farmland's natural resource base is therefore weakening, while our production and profit margins become less reliable.

Unfortunately, the sustainability issue is not limited to agriculture. We are violating the major principles of sustainability in the natural, human, and economic activities of all aspects of our society on a global level. We're extracting reserves of fossil fuels, minerals, and metals faster than they can be re-deposited and reintegrated. We are harvesting, consuming, and disrupting our ecosystems (such as rainforests, fisheries, waterways, oceans, aquifers, and farmlands) in ways that diminish their productivity and at levels beyond their capacity to maintain. Our waste and man-made materials are being produced at a faster pace than they can be broken down and integrated back into the cycles of nature, or deposited into the earth's crust and turned back into nature's building blocks. Only the most adamant deniers disagree that change is urgent, and most agree that we must make profound changes in the near future to avoid a severe breakdown at the biosphere level.

But, there is hope. We can still turn these conditions around. Fortunately, we have pockets of dedicated advocates, grassroots organizations, and practitioners throughout the world pointing the way back to nature and sustainable living. But we must understand the roots of what we are facing. The fact is that across all industries and walks of life, we have a severely diminished practical understanding of nature and its crucial role in every aspect of our lives; and we have lost our spiritual connection with her. As our daily interaction with the natural world has diminished, so has our awareness of how we are totally reliant on nature and our understanding of how she works. As the bulk of our population has moved from an agrarian society to an urban one, we have lost our connection to the land. We no longer identify with the two unifying forces—how every bite of food we eat and every ounce of air we breathe are dependent on the healthy capture and transfer of energy from the sun by green plants, and the proper cycle of essential materials. But as the health of these processes deteriorates, so does the availability and quality of food we eat and the air we breathe.

Yet, that is just the beginning. As important as the role of nature is in providing for these tangible needs, our greatest loss is going on inside us. It's our relationship with nature, our direct ongoing exposure to her grandeur, beauty, biological and physical marvels, her mighty rush of power in the midst of turmoil, and display of serenity in time of calmness—it's these qualities that fuel our inner being and cause our souls to sing and spirits to dance. Anyone who has ever strolled along a country path at the break of dawn and seen the sparkle of dew on the plants, noticed the glistening wings of a dragonfly darting past, heard the chirping of birds and the rustle of squirrels, smelled the fresh scent of dogwood blossoms, knows of what I speak. With these interactions we experience a certain oneness with the universe and our inner being is renewed. Studies have shown that this stuff is real and affects our behavior and health in measurable ways. Hospital patients do better when they have access to nature. Without these encounters and a sustained connection with nature, that same spirit begins to wither, and something within us begins to die. In the words of Norman Cousins, "Death is not the greatest loss in life. The greatest loss is what dies inside us while we live."

I've seen this growing disconnect with nature firsthand, in ways that have been instructive, if not painful. Perhaps, because of my background and career, I have apprehended in a more personal way our inextricable link to nature and our dependence upon it. As you will read in the following pages, I learned of this link as a little boy in rural Mississippi, where I lived close to nature and explored her wonders in an environment unadulterated by modern technology. Then, as a research scientist, I have continued to delve into the richness of nature. I was blessed to work shoulder to shoulder with outstanding colleagues and brilliant, inquisitive young scientists from around the world, and we continually shared our admiration for the ever-unfolding elegance within the natural systems we studied.

I have also spoken extensively to audiences from all walks of life about our discoveries—students of all ages from grade schools through colleges, civic clubs, professional groups, church groups, and people of all ages and educational levels. I talk about farms and farming. But not about big tractors and deep tillage, or about pesticides, fertilizers, and other inputs and interventions. I don't speak about huge monoculture fields and sixteen-row planters, nor the latest in precision agriculture methods, or GMOs, or any of the other multitudes of modern technologies that have been ushered in to ostensibly "feed the world." No, I talk to them about farming of a very different kind, using a different farm language and with a different farm voice. I speak about our role in the remarkable web of life, and the power of natural solutions, and how, if applied wisely, these solutions will provide abundantly for us and generations to come.

Without fail, something within them resonates in a very personal way as I share the discoveries about smart wasps and talking plants—how plants under attack by caterpillars send distress signals to their bodyguard wasps; how insects learn and can be trained to detect explosives in a similar way as bomb-sniffing dogs. I explain how these amazing findings demonstrate built-in strengths within crops that are more powerful and safer than any pesticide, and how we continue to work with on-farm studies with growers, successfully demonstrating how these systems work as well or better than conventional methods, and with better ecological and economic benefits. I stress that we have the opportunity to reclaim our true heritage with these practices. But the hour is late. We must move swiftly as a society to reconnect with nature and regain our "farm language" voice.

In these settings I always see an interest in nature, a desire to reconnect with her. Though weakened, it's still alive. The embers are still glowing. And maybe it was the satisfaction of the Wolf Prize, or perhaps just the buzz of the coffee, but the thought of that small glow gave me hope that morning as we circled above the Atlanta airport. It made me think of what Marilyn Ferguson wrote in her book, *The Aquarian Conspiracy:* "We know that the very forces that have brought us to planetary brinksmanship carry in them the seeds for renewal." I have learned the truth of this. In my journey from backwoods to modern science, I have learned this and more.

CHAPTER 1

Simple Life in Simple Times

"Hold still, Wallace Joe."

I can still hear my mother's command as she stroked a comb through my hair. I was four. Patsy, my sister, was sixteen months older than me and Mother was getting us both ready to go to the cotton field. Day was only just breaking, casting early-morning light into the room. It was another muggy July morning in Mississippi.

Normally, Mother just called me Joe, but if she started with my proper first name, that meant business. So, I held still and let her finish with the comb. Then she made sure my homemade shirt and short pants were properly aligned. I never could decipher why you needed to comb your hair or be so neat to work in the field, but those were the rules in our house. At least I didn't have to wear shoes. Children in my little corner of the world didn't wear shoes in the summer, except to church on Sunday, of course. The end result was that the soles of your feet toughened up pretty good in the summer months and you could walk and even run on most any surface, including the gravel roads that connected our community. And although I had to wear it out the door, I could take my shirt off for most of the day, too. A pair of shorts was all a child of four in Mississippi needed.

"Your Daddy's already harnessed Ole Nell," Mother said. "He's probably in the field plowing by now. We want to finish our work in the cotton field this morning so we can work in the garden this afternoon."

Daddy and "Ole Nell" plowing.

Mother was a large, robust woman, but seemed to glide swiftly and tirelessly through her many chores. I can still see her with the ever-present hand towel around her neck, to wipe the perspiration from her face or to swat at the house flies, which had unrestricted entry through the open and screenless windows. As usual, Mother had risen well before dawn to build a fire in our wood cooking stove and then to start preparing breakfast—biscuits, eggs, sausage, and cane syrup. Every morning she'd wake us when breakfast was ready and every morning we'd eat together as a family. Afterwards, she'd handwash the dishes and start the initial preparations for the dinner meal, which we'd eat around noon. The midday meal was "dinner" in those parts, the evening meal being referred to as "supper."

Daddy was up at least as early as Mother. Before breakfast, he'd milked the cows, fed and harnessed the mule for the field, and brought in additional wood for the stove. The house was warm and stuffy from the cooking, so we were all eager to get outside and hopefully catch the occasional early morning breeze.

Our residence was an unpainted, three-room shack with a tin roof and a small front porch. My parents didn't own it. Our family was one of several sharecropper tenants on the land of a country doctor. One of the two bedrooms had a fireplace and a small sitting area that opened to our tiny kitchen with an adjoining dining area that had just enough room for that wood burning stove and two tables, one table for cooking and washing dishes, and one with four chairs for dining. An exhaust pipe led from the stove to an outlet in the side wall, under which an ongoing supply of stove wood was stored. Other than two windows at the back of the kitchen, the house had only wooden shutters, which could be opened when desired for light and air.

We had no electricity or running water. A kerosene lamp was the primary light source and we always kept a second lamp in reserve. For outdoor activities requiring light, we had a portable kerosene lantern. With only one fireplace along with the wood-cooking stove, winters were cold and at times downright bitter. With no air-conditioning or even the presence of electrical fans, summers were hot and sometimes brutal.

Every drop of water we consumed had to be drawn, bucket-by-bucket, from a well located at the back steps. The galvanized bucket was attached to a rope threaded through a pulley and wrapped around a wooden reel. The reel turned by way of a handle and you simply lowered the bucket and reeled it back up. The well was sheltered by a tin-covered canopy and was furnished with a shelf where a bucket of fresh water was always present, along with a dipper and a pail for washing the face and hands. A one-seater outhouse rested about 50 feet from the back door, which seemed like 150 feet on cold mornings. Toilet paper was a luxury and we made do mostly with the pages from the Sears & Roebuck catalog.

We had a smoke house and a chicken house and farther to the back was a small barn equipped with a couple of stables and corn crib. But it was the cotton field that drove the farm— five acres that provided the lion's share of our annual income. And on that summer morning, as most summer mornings during growing season, Mother finished our grooming, donned her wide-brimmed straw hat, and off we went to the field. Mother stopped near the back steps to grab our wooden wheel- barrow having already loaded it with her hoe, a bed sheet, a gallon jug of fresh water, and a pint jar with lid.

I carried one of my favorite toys, a three-foot stick with the lid of a molasses can nailed loosely to one end so that it would turn like a wheel. I'd push it in front of me imagining it to be a big truck or some such powerful vehicle. In my pocket was a corn cob with a couple of chicken feathers poked in one end to act as rudders to give it something of a dart effect when I'd throw it at various targets. Patsy was carrying a homemade cotton-stuffed doll that stood in sufficiently for the Raggedy Ann doll we could not afford. She had a small sackcloth bag, too, that held various lids and similar objects that she'd use for making mud pies or as kitchenware for playing house.

Mother Patsy and I going to the cotton field on an early July morning.

We walked briskly down a path and through a thin line of trees to the cotton field on the other side. As we passed through the tree line, I caught the scent of honeysuckle mixed with the odor of fresh-tilled soil, and I could hear the chatter of squirrels rustling nearby. It was the dawn of a new day and all about, life was swinging into action. Soon I could see Daddy and Ole Nell against the morning skyline, moving steadily across the field. This was a sight I would witness many times, a scene that would become deeply etched in my mind, the prototypical farming scene of the era—the beast of burden in harness, diligently and obediently pulling the plow; my father, the master, skillfully steering the plow along each row, calling out the commands: "Gitup" for go, "Whoa" for stop, "Gee" for right, and "Haw" for left. Man and animal moving along in perfect cadence.

Daddy was wearing his overalls, light blue work shirt, brogan shoes, and a straw hat. As an avid smoker, he'd periodically stop, usually at the end of a row, and slide a can of Prince Albert tobacco out of the front pocket of his overalls, along with a packet of rolling paper. He'd niftily roll a cigarette, light it with the lighter he kept in his side pocket, and start plowing again, now puffing on the cigarette as he and Ole Nell would begin a new row.

We entered the field and Mother promptly selected a spot near one end where she would soon begin her work cleaning out clusters of Johnson grass and other patches of weeds difficult to remove by plow. There, she spread the bedsheet over the top of a couple of the rows of cotton plants, providing Patsy and me a shaded area between rows where we could sit and play. This served as a home base and we were expected to stay reasonably close to it, at least in the general view and easy calling distance of Mother. Through the morning, as her work shifted, she'd occasionally relocate the sheet to a position closer to her. She'd find an especially shaded area for the water jug, partially burying it in the cool soil with the pint jar beside it.

From time to time, I could expect Mother or Daddy to holler, "Waterboy!" upon which I had to fill the jar from the jug and run it out to them.

In these earliest days of my childhood, I no more noticed the wildlife around me than the soil, or the cotton, or the trees and sky. That is to say, it was just a part of the landscape, a part of life itself. But as I would grow, these visits to the field yielded closer observation. I would become fascinated by the insects, birds, and other animals that made their homes on that land. Dragonflies—we knew them as "mosquito hawks"—of several different colors, shapes, and sizes were abundant. Some would hover close to the cotton plants, often landing and sitting for a bit. Others flew higher and seldom landed. Daddy would show me, when a mosquito hawk was perched on a plant, how you could sometimes catch it by slowly creeping up from behind and, with a steady, easy-moving hand, nab it by the tail (which I would later learn was actually its abdomen) between your thumb and forefinger. I'd find them elusive and would continually work to improve my technique. Upon an occasional capture, I'd hold that mosquito hawk for a few minutes for a closer examination, then release it and watch it zoom away.

Daddy would also teach me how to "fish" for doodlebugs. A doodlebug was just a type of worm to me then. Years later, I would learn that these critters are the larval stage of the tiger beetle. They have armored heads with a claw-like apparatus as their mouthparts with which they snare and feed on ants and other small organisms. They lived down the small, circular holes that we saw around the edge of the cotton field, holes of a diameter roughly that of a pencil. Daddy would show me how you could take a piece of straw or a pine needle and insert it into one of those holes and make a small poking motion. Now, while you were doing this, it was important to chant, "Doodle, doodle, your house is on fire!" Soon enough, you'd feel a small tug on the straw and by pulling upwards at just that moment, you could snatch the doodlebug right out of his hole. I'd do this endlessly, inspecting the displaced creatures and safely returning them to their lodgings. By introducing the tip of a captive doodlebug's tail into the hole, and with a bit of encouragement, it would back snuggly into its burrow, though no doubt stunned for a bit. (As I'd get older, it would become apparent to me that Daddy's "house-is-on-fire" chant was a step that was purely optional.)

As the sun was approaching its midday high on that summer day when I was four, Daddy came over to us to tell us it was dinner time. "I'm going to water and feed the mule and will meet y'all at the house." It was standard practice to take the mule to the small pond behind the barn for water and to feed her a few ears of corn during lunch break. Daddy would also take the harness off to let Ole Nell cool down before putting her back to work after dinner.

Mother gathered up her supplies and loaded them into the wheelbarrow and we started back to the house, the noontime sun now beating down on us. The soil on the path was now warm on my feet. There was more of a stillness in the air as compared to the buzzing of activity I had noticed through the tree line at dawn. I trotted out ahead, pushing along my molasses-lid toy. Recently, I'd learned to draw water, so my task was to pull three buckets of it. One was placed on the shelf at the well while the other two were for kitchen use.

We all arrived at the house hot and fairly dripping with sweat. The first step was almost ritualistic and of immense pleasure. One by one, we each gulped down a dipper full of the cool, freshly drawn water, and then each poured a second full dipper into the wash pail. By cupping both hands together the water could be scooped up and used to thoroughly wet and wash the face and neck, refreshing us greatly in the process.

With Patsy's assistance, Mother began preparing the table and food for dinner. The meal consisted of garden fresh butterbeans, squash, sliced tomatoes, turnip greens, and cornbread, along with sausages left over from breakfast. The cooking had been done that morning, along with breakfast, and the items had been kept warm in the warming portion of the stove by the hot coals left in the firebox compartment. After eating, Mother quickly cleaned the table and washed the dishes in the dish pan and Patsy towel dried them.

Truth be told, there had been nothing special about this particular morning. Or about the afternoon that followed. And yet, in a sense, there had been everything special about it, though it would be years before I would understand that. It all meant something. Every activity held its own share of significance. That afternoon, we probably worked in the garden. Maybe we churned some butter. Many times I was given the assignment of taking a jar, filling it halfway with milk, and persistently shaking it for twenty to thirty minutes. "Come, butter," I'd repeat the whole time (for no better reason than telling that doodlebug his house was on fire). The end result, after you poured out the remaining milk was the creamiest butter a person could ever hope for.

Maybe it was wash day. These were typically Mondays. We'd build a fire under a big cast-iron pot filled with water into which we'd throw all our clothes, along with homemade lye soap. Mother would stir that pot as it boiled, then take the clothes out and put them into one galvanized tub to further scrub them, and then another tub to rinse them. Then they'd get hung up on the clothesline to dry.

I'm sure there was some wood-cutting that afternoon. We probably went down to the river to bathe and cool off. If it rained, Mother would often make teacakes.

In the early evening, before the night would close in, it was time for supper. When darkness finally fell, it always seemed to come with just a touch of melancholy for me. The kerosene lamp was never sufficient to stave off the inevitable ending of the day. I'd find my daddy by the light of his rolled cigarette

Enjoying the security of Daddy's lap during the cloak of darkness.

and sit in his lap, feeling safe and secure once again in his rough, yet gentle, hands. Mother would read by the light of the lamp. Often one of them would tell Patsy and me the story of the "Three Little Pigs" or "Little Red Riding Hood" or one of many other such stories—our evening's entertainment.

Where and when I grew up was, in other words, a simple place in a simple time. Our little nuclear family was held closely together in tight quarters by love and duty, providing me with a sense of comfort and belonging. Our lives were uncluttered. We were unencumbered by the lure of materialism. By any standard of today, my childhood was impoverished. We had no electronic games or gadgets; no radio, television, or electronic communication of any kind. Patsy and I played with our improvised homemade toys. And we wanted for nothing.

It was, to quote Dickens, the best of times and the worst of times. Our subsistence was forever at a minimum. We had limited shelter from the heat and cold. Our food supply was based directly on what we produced with our sustained hard labor. Yet the stark simplicity of our lives promoted a connection with each other and—perhaps the most important of our connections, at least where it came to our daily survival—with nature. And that relation bred an appreciation for its beauty and grandeur. It was omnipresent. I remember well waking to the chirps of a cardinal followed by the sound of the mockingbird and the crowing of the roosters. This repertoire of nature's music continued unabated throughout the day, ending with the sounds of the whip-poor-will in the evening and the hoot owl at night. The profound absence of electric lights necessitated a linkage of our daily activities with the natural cycle of daylight and nighttime. Without the multitude of distractions associated with commercial and industrial commotions, we were keenly aware of the sounds, smells, and sights of nature and the rhythm of her daily and seasonal patterns. Our senses and emotions were in tune with this language of nature, and we instinctively learned to read it proficiently. We could relate it to the time of day, the coming and going of seasons, the approaching of storms, key information about the health of the environment, and a vast array of other such information.

Our relationship with nature was, indeed, more than just a connection. Looking back, I can see that we were not only connected to her, we were of her. Our family was an integrated piece of a larger, holistic structure, a web of interwoven parts that was reflected in the way we not only farmed, but how we lived and even how we related to one another. Our family was a microcosm of the larger whole, each piece—father, mother, daughter, son—forming an interconnected system. There was little within our family, or farm, or the natural world in which both functioned, that existed independently. I didn't see this at the time of course because I was a part of it. I was inside of it. Only much later, reflecting on these days and comparing and contrasting them with the world of the late twentieth and early twenty-first centuries, would I see the difference, the movement away from this web of nature, the separating of the parts, the propensity for society, via technology, to not so much work with nature as to *intervene* in nature.

But these were thoughts that were years away from the mind of a young boy on his daddy's lap listening to the tale of "Little Red Riding Hood." Soon, it would be time for bed and another day in Mississippi would pass, ordinary and extraordinary all at the same time. I would never have guessed the latter description back then. How could I have known the personal meaning those days would have? And never in a million years of Mississippi summers could I have envisioned the professional meaning. All of that was ahead of me. Impossibly far ahead.

CHAPTER 2

When Community was Community

When my mother went into labor to deliver me on October 30, 1942, my father had to fetch the doctor on foot. In the rain, I am told. My parents had no other means of transportation and wouldn't for another ten years when, finally, a secondhand Ford Model A would take its place in front of our home. The doctor delivered me in my parents' bed, just as he had delivered Patsy in 1941.

Patsy was born a few months before the Japanese attack on Pearl Harbor and the country was fully engaged in the war by the time I came along. I don't remember much if anything of the Second World War, but I imagine that life probably only got harder for my parents in those years, both of them having spent their early adulthood surviving the Great Depression. The postwar boom would miss most of southwest Mississippi, where very little industry was developing. The economy continued to be centered on small family farms that mixed cotton and corn with small beef and dairy operations. Cotton was typically the cash crop with corn being grown mainly for feed for the family livestock—milk, cows, chickens, swine, and work mules. For the most part, farms were either one-mule or two-mule farms. Ours was a one-mule farm, that mule being Ole Nell. Milk, eggs, beef, and pork came to the table directly from the family livestock, with vegetables coming from the garden. You ate what you raised on your farm, in other words, with the sale of your cotton providing for all your other needs.

My mother, Grace Carruth (with no middle name), was born on September 28, 1917 in north Amite County, Mississippi, to a large rural family with a total of ten children. My father, Ferd Arthur Lewis was born on February 5, 1911 to a family just as large (also ten children) about ten miles to the north in south Lincoln County. They married in February of 1940. Ferd, who took his unusual first name from his maternal grandfather, developed his farming abilities early, along with hunting and fishing skills. He saw no real need for school and dropped out with only a rudimentary knowledge of arithmetic and no ability whatsoever to read or write. Grace, on the other hand, boasted a high school education and was an avid reader her whole life, finding topics on medical and health sciences especially interesting.

When they married, my parents embarked on their life as sharecroppers, living and working on available farmland. They moved around an area west of McComb no greater than five square miles. Their requirements were the use of the land, plus a tenant house in which to reside. In return, they'd typically give a fourth to a half of the profits. That tenant house we lived in when I was four was one of several we lived in during my childhood, three of them on the same landlord's property.

That landlord was Marion DeKalb Ratcliff, M.D., the only doctor I knew throughout my childhood. (I have no recollection of Dr. Walker, the doctor who delivered me on that rainy night in 1942.) Dr. Ratcliff lost a leg in a streetcar accident at a young age and moved about quite nimbly on crutches. He made house calls as needed, driving an old Plymouth with a throttle that he used frequently to assist him with his disability. The doctor's liberal use of the throttle became a familiar sound that signaled his coming and going to the entire community. There was a small hospital in the town of McComb, but it was about ten miles away and was difficult to get to. Nobody went there except in extreme emergencies. And so to me and to many of the kids around the area, this country physician was our only exposure to a doctor of any sort. The mother of my six-year-old cousin once admonished him to study hard so he could one day become a doctor. "I wouldn't mind being a doctor," he replied thoughtfully, "but I'd sure hate to have one of my legs cut off."

Almost all the roads in the communities of southwest Mississippi were dirt and/or gravel. Cars were scarce. Many people walked, and in addition to the roads, a network of pathways established by repeated use provided shortcuts through the woods and fields to neighbors and other locations of central interest, reflective of the community's ongoing social linkage. Each house received RFD daily mail service to a mailbox placed at a nearby roadside site. An ice truck often came through delivering blocks of ice for use in the iceboxes of the day, the forerunners of electric refrigerators. The practice of sending written notes or verbal messages via the mail carrier or the ice deliveryman was frequently used as a means of communication.

The worn paths led to one particular site of major interest—the Crossroads Grocery Store, owned by my father's older brother, my Uncle Lonnie. Unlike my father, Lonnie could read and write, at least at a basic level. He and Daddy were spared the war, but two other brothers were called into it, one to the European theater and one to the Pacific. Both these uncles of mine came home physically intact, though one had frostbite damage to his feet while the other carried some shrapnel in his hip.

The store was the area's social gathering spot, especially when farmers were "laid by," meaning that the cultivation activities were completed and they were waiting for harvest time of

Uncle Lonnie's Grocery store, the community gathering spot.

their crops. People would sit on the store's front porch on benches, chairs, and the front edge of the porch floor, legs dangling, and chat about all manner of topics, from their crops, to the weather, to politics, and, naturally, to the latest gossip about other people. Often, we'd pass away the time there until dark, having to make our way home along the paths by lantern light.

Uncle Lonnie was somewhat of a jokester and often had something lively and comical going on at the store. One time he had an object that when placed strategically on the porch floor or steps looked precisely as though a dog or cat had defecated there. We spent a delightful portion of the day sitting around watching and laughing at people's reactions as they came upon the bogus sight in their paths.

My grandfather Lewis—Joe Reuben Lewis, after whom I was named and whom we all referred to as Pap Paw—lived a short walking distance from the store, and Pap Paw spent a major part of most days there in a rather patriarchal position. He always sat in a special chair in a particular place—a straight chair, tilted back against the wall on its two back legs. Pap Paw spoke frankly, but in a gentle manner, particularly to all his grandchildren when they came around.

Other social gathering spots included the churches of the area. The Lewis family attended Mount Gilead, a Primitive Baptist Church where my great uncle was preacher. His father had been the preacher before him, and his nephew would be the preacher after him, both of the men for around forty years each. The church had no baptismal and I was baptized in a nearby creek. Services were not every Sunday. Typically they were held a couple of times a month, but what was lost in frequency was made up for in intensity. These services were all-day affairs, which included "dinner on the ground." There would be a morning service, followed by a lunch on picnic tables outside, and then an afternoon service. The afternoon service usually included preaching but sometimes consisted only of the singing of hymns. There was no piano or accompanying instruments of any kind. All the singing was acapella with a number of outstanding singers present for the various musical parts. The harmony of all of our voices blended into a single enormous voice, it seemed to me, with "Precious Memories," "Jesus Paid it All," "Amazing Grace," and "Sweet By and By" resonating powerfully throughout those country hills. The sounds and lyrics of those old songs are spiritually moving to me even to this day.

Following that afternoon service, we'd often retire to the Crossroads Grocery where other families might be idling the rest of the Sunday away. Everybody knew everybody. Some of the residents of the area came from families that had made their homes there for three or more generations. Most were farmers and they were people you could count on if you needed help, though nobody seemed to really ask for it. I remember the characters that came and went from that store. Mr. Freeman, always strongly squinting one eye, drove a gravel truck which, like Dr. Ratcliff's Plymouth, you could hear long before its arrival. Pert Johnson was hard of hearing and we kids were perpetually amused by how loud he talked. Then there was Clarence East, who always had a cheek full of chewing tobacco with a bit drooling down the corner of his mouth.

Of course, the store was also a gathering spot for the children of the area and we'd often find ourselves engaged in all types of games. In the earliest years it might have been "hide and seek" or "ring around the rosey," with different generations often joining in. Later we moved to adjoining fields for more serious levels of baseball and basketball, with the role of the adults becoming more advisory. Sometimes they'd volunteer to be umpires or referees. I enjoyed these sports and was blessed to have a number of cousins to play with. Uncle Lonnie had two children, Paul and Charolette. Daddy had two sisters who lived in the community, one with three sons—Charles, Robert, and Leslie—and the other with a son and daughter—Gabriel and Deloris. Paul, Robert, Leslie, Gabriel, and I were all within two years of each other and we seemed to share natural athletic abilities. The five of us, along with other youngsters in the community, spent many wonderful hours honing our baseball and basketball skills in those adjoining fields.

One year, someone from Brookhaven, Mississippi recruited all five of us to play together on a Little League team. To the chagrin of some, we "cousins from the county" dominated the league that season, winning every game. The bonds forged in those days would continue well into later years. During my senior year at West Lincoln High School, I would be the catcher and Leslie the lead pitcher of our baseball team. We would make it into the final four of the state championship that year. I'd be recruited to play American Legion baseball and be offered a baseball scholarship to junior college, though opportunities for academic scholarships would ultimately lead me elsewhere. Meanwhile, Robert and Gabriel would go on to have successful careers as high school coaches.

Besides churches and the store and various sporting activities, community socializing also presented itself on simple evening visits to and from neighbors. These visits typically took place on someone's porch. It's what porches seemed to be for, so far as I could tell. Of course the adults would talk about grownup subjects that none of us children had any interest in. We preferred spending those late evenings catching lightning bugs. Every now and again, however, someone would capture our attention with a ghost story or two. Like most kids, I had a love-hate interest in those stories. There was an irresistible appeal that nevertheless always resulted in spooky feelings that would take me a while to shake.

Whatever the means, we were, all of us in the community, connected in a time and place that one doesn't ordinarily think of when one thinks of connectivity, at least by today's definitions. I had no explicit awareness of this connection at the time, of course, for I knew nothing else. But I was linked to those people—the kids, the members at Mount Gilead, Dr. Ratcliff with his one leg, Mr. Freeman with his squint, Pert Johnson with his loud talking, Clarence East with his cheek full of tobacco, others that gathered at the Crossroads Grocery, and everyone else with all of their own little idiosyncrasies. Just like our family's farm and the natural world around us, the community, too, represented a larger whole of which we were an intertwined part. Only later in life would I see this relationship for what it was, mainly due to its absence, lost as it would become to the unrelenting progress of the years: Yes, I was a part of that community. But that community was also a part of me.

CHAPTER 3

Cotton, Corn, and a Mule Named Nell

The sun was low in the west, my legs were aching, and I was hot and drenched in sweat. But I was bursting with pride. At seven and one-half years of age, I had arrived at the notion that I was on the threshold of manhood. The mule and plow had been under my management for the full day, and when I hit the end of the row I was plowing, that meant I had single-handedly plowed our complete five-acre cotton crop.

Diligent, year-round work by the entire family was a necessary practice of farm life in the community and children were expected to move into their roles at early ages. As a typical youngster, I could hardly wait to be a grown-up, and the real mark of manhood for young males in our community was the independent handling of a mule and plow. I had become tall enough now to reach the handlebars of the plow, and Daddy had taken notice. The day before, he'd spoken the words my impetuous youth had yearned to hear: "Son, you need to get a good night's sleep because it's time for you to take over some of the plowing. Tomorrow, I want you to sweep the middles of the cotton field."

I hardly slept that night. Early in the pre-dawn, while Mother was cooking breakfast and Daddy was milking, I took Ole Nell to the pond to get water and I placed her in harness. Although I had done so numerous times before, that morning I applied the harness with a renewed sense of care and conviction. I returned to the kitchen and quickly downed a hearty breakfast and then headed for the small tool shed near the barn where I had tied Nell to a post. The "middle sweep plow" was waiting there and I had already mounted the singletree to the front end of the plow.[1]

"Whoa back! Whoa back!" I commanded, using the plow lines to draw Nell backward so the trace chains from her harness could be hooked to the singletree. The initial rays of morning light were just radiating upward along the eastern horizon as we made our way around the edge of the field to the starting location, with the plow expertly tilted on its side and guided by one handlebar.

"Git-up Nell, we have a full day's work to do," I commanded with every bit of seven-year-old authority I could muster as I set the plow point into the first furrow. It didn't take long before I could see that this manhood role of plowing was an exceptionally hard and rugged job. That day marked a powerful beginning to an enormous education. The "one row at a time"

---

1 A singletree is a wooden or metal bar that is hinged at its center to the plow with hooks on each end where the trace chains from each side of the mule are connected. The middle sweep plow had a V-shaped blade with a somewhat rounded point with flanges reaching out equal distance on either side. This blade was mounted at the bottom of the plow stock at a gently downward angle with the point being the center, lowest, and leading part of the blade. The plow was pulled along the center of the furrows (usually about thirty inches wide) running between the rows of plants growing from the top of bedded rows on either side of the furrows. The plow blade, which was near the full width of the furrow, would slice a couple of inches deep, and in the process remove weeds and toss fresh loosened soil toward the bed on either side. Other plows and devices such as the turning plow, side harrow, and middle splitter, with different designs and for other purposes were used at varying times of the seasons, but were pulled by the mule, operated and guided in the same general way.

My first full day of being in charge of plowing "Ole Nell"
at seven and one-half years of age.

pace began to set in, and the field looked bigger and bigger. All
told, I would end up plowing fifteen miles that day. Plowing,
I would go on to learn firsthand, consists of long, lonely days
often walking twenty to thirty miles a day under the same harsh
conditions. And there were various hazards to the work with
which I would soon become thoroughly acquainted. Hours of
walking behind the plow means you become intimately familiar
with the anatomy and functional activity of a mule's hind end.
Because of my experience, I can fully relate to the saying that
"if you ain't the lead dog, the scenery never changes." And the
scenery wasn't pretty. The mule naturally has periodic episodes
of defecation, urination, and flatulence. There are, therefore,
significant olfactory consequences that are frequently presented, along with perilous walking pitfalls. Fortunately, I quickly
learned the signals that telegraph each and became quite adept
at minimizing the hazards associated with these occurrences.

In truth, I had been practicing plowing under my father's
supervision for weeks prior. For over a year, I'd been putting
on the harness and hooking up the plows. But in leading to my
first day behind the mule, the instructions had become more advanced. They were clear and emphatic and delivered in the way
only Daddy could convey them: "Ya hafta learn how to work

with the plow and you hafta to learn how to work with the mule. The work's gotta be smooth and straight, always the right distance from the plants. Ya won't amount to nothin' if ya can't plow a straight row. Now, don't fight the plow. Let the plow do the work with easy movements. If ya rastle with it, it'll work ya to death and yer plowin' will be ragged. And ya hafta talk right to the mule. Ya gotta let her know yer the boss. But don't holler add'er. And never beat'er. If she acts up, just tell'er 'no!' with a firm tap with the plow line. If ya let her know what ya want and are good to her, she'll plumb work her heart out fer ya."

The age-old practice of plowing with a mule represents one of humankind's earliest successes in using nature for our long-term survival. As a cross between a male donkey (jack) and a female horse (mare), the mule combines the endurance and sure-footedness of the former with the size and power of the latter. The mule's recorded use as a valuable beast of burden reaches back to biblical times and includes instances of pulling and carrying loads for warfare, construction, farming, and many other purposes. There's a particularly rich history of the mule with farming in the Southern United States. Almost all farms were dependent on their service for growing crops up through the 1940s and 1950s. Methods and processes for breeding, training, and handling of mules and the design and handling of plows and other equipment co-evolved and became imbedded extensively in the Southern practice of farming.

In my own experience over the next several years on our farm, I began to see more and more the beauty in Nell's plowing performance, and to further appreciate the art of the plowing operation, built as it was on the vast accumulation of methodology reaching back to those ancient times. At the center was an interactive performance of the mule and operator, a sophisticated dance between Nell and me involving ongoing signals, feedback, and synchronized responses. I noted how keenly Nell heeded the verbal commands, how she would adjust immediately to the feel of my left and right movements of the plow, always maintaining a proper location in relation to the row of plants. She would keep a steady speed and gait by making continual adjustments in her "power of pull" in relation to the firmness of soil and depth of the plow. If I lifted the handlebars of the plow to guide the plow point deeper, I could feel Nell immediately strengthen her pull to retain her speed, and vice versa if I pressed down to plow in a more shallow way. Through these coordinated movements, together with numerous other such timely interactions, we were able to move along in what felt like a gliding flow. It's the well-trained mule that makes the job tolerable versus impossible. Those who have experienced a plow handle repeatedly hit them in the gut due to a mule that fails to walk evenly or that abruptly slows the pace knows this difference very well. More importantly, this successful dance between man and beast provided for quality plowing in terms of such vital considerations as straightness, evenness, and uniformity of depth and width of furrow.

It would be a day in March every year that the growing season would get underway. I'd come home from school one afternoon to the smell of freshly tilled soil and would know that Daddy had begun breaking the ground for our crops. Although certain livestock chores had to be maintained year-round, the central focus during the growing season was the cotton, corn, and garden crops. Once that magic day came around, I knew that all the days would now be long and hard, and that I'd need to go straight to the field upon returning from school.

I'd change into my field clothes and head out to meet Daddy, typically stopping by the food safe to get my usual after-school snack—a leftover biscuit into which I would poke a hole with my finger and pour in some molasses. I'd then take over the plowing, leaving my father free to attend to any one of the other endless chores that needed doing, maybe mending a fence somewhere. Often Willie, a neighbor, would come over and help my father as Daddy and Willie often "swapped work," a common practice in our community.

Besides the five acres of cotton, we had ten to twelve acres of corn. For both, we'd start by using the turning plow to break the land. After completion of breaking, we'd come back with the same turning plow and "row it up"—prepare the rows of seed beds where the seeds would be planted. The turning plow was a heavy pull and some folks used two mules for this. Though Ole Nell was a relatively small mule, she was strong and could handle the turning plow, but not at a fast speed, especially for several days in a row.

After rowing it up, there was the planting, which required the use of an old one-row planter, also pulled by Nell. Then each field had to be cultivated three to four times, row by row, by mule and plow using a combination of side harrow, middle sweep, and turning plow. Further, each field had to be cleaned of certain grasses and weeds with a hoe at least once. At harvest time, typically around September and October, the corn was hand-pulled ear by ear and the cotton was hand-picked boll by boll. With nothing beyond one mule and hand tools to accomplish everything, there was little time to spare. It was of great benefit to Daddy when I grew old enough to work Ole Nell. Daddy and I did all the plowing while Mother and Patsy helped with the hoeing and cotton picking.

We had some help in the cotton fields by a flock of Cotton Patch geese, a breed of goose that fed off of the grasses that inevitably sprouted up around the cotton plants. Both broadleaf weeds and grasses would grow throughout the field. The geese fed only on the grass, so they offered no threat to the cotton plants. So, by having the geese maintain the field clear of the perennial grass, it was much quicker and easier to go through with a hoe and chop out the broadleaf weeds, which were annuals. In an age before herbicides, our task was made much more simple by the introduction of the geese—a natural solution to a natural problem. Of course we didn't know of any other solutions.

We also had another menace: the notorious boll weevil, *Anthonomus grandis* Boheman. Migrating north from its native Mexico and Central America, it had entered Texas some fifty years earlier and marched into cotton fields throughout the cotton belt to become a major pest with a sweeping impact on the cotton industry and, indeed, the entire economy of the South. Shortly after World War II, toxaphene, an insecticide related to DDT, became available and provided reasonably good control of the boll weevil during the time we were farming. I remember occasions in late afternoons when the air was still; Daddy—dressed in old overalls, a long-sleeve shirt, gloves, and with a handkerchief around his face—would dust the field of cotton with toxaphene. He would walk briskly through the field shaking a croker sack of the chemical dust over the tops of the plants as he passed by. Later on, as I learned more about the risks, I found myself a little surprised that no subsequent health consequences of this exposure ever became apparent during Daddy's ninety-three years of life. In any event, the boll weevil began developing resistance to toxaphene and related insecticides by the 1950s. At that time, a new family of insecticides called organophosphates became available.

There was much at stake with our cotton field. Maybe everything. As was the case in most of the South, cotton was king in our community, even with sharecroppers. Our five-acre cotton crop provided the primary source of money, limited as it was, for all purchases throughout the year. These included farming resources such as seeds, fertilizers, and small farm tools like hoes and plow points. From this meager income came all of our personal items, too. But this was only after one-fourth to one-half of the income, depending on varying arrangements, would go to the landlord as payment for the house to live in and use of the land. The return on our cotton crop determined the ability to negotiate favorable sharecropping terms as well as our all-around monetary well-being for the year. There was no room for inefficiency or downtime. We had no bank accounts or reserves and certainly no property or any other type of equity. A few head of livestock, a little furniture, and the clothes on our back represented our collective net worth.

The ten to twelve acres of corn served primarily as feed for the livestock—chickens, cattle, Ole Nell, and, on occasion, some swine. Corn was ground into cornmeal for cooking purposes, too, with cornbread being one of our regular foods. Occasionally, enough corn was available to sell as a supplement to the cotton income. Throughout the growing season, the garden provided a variety of fresh vegetables for the table, including tomatoes, okra, butterbeans, peas, turnip greens, mustard greens, potatoes, and watermelons. In the absence of refrigeration, it was important to regularly can some of the vegetables for use in the winter months, so canning was a key activity at least a couple of times during this period. Other foods consisted of milk, eggs, and meat produced directly on our farm or harvested by hunting and fishing. We made use of the ancient practice of gathering, too, picking fruits like blackberries,

Daddy milking "Ole Jersey," one of our two milk cows,
a chore that we all had to learn.

plums, figs, or pears. Very little food was purchased.

Management of our livestock required attention at some level on a daily basis year round. We maintained a small herd of eight to ten cows, including calves and yearlings. Two to three cows served as primary milk cows with milking required in the morning and late afternoon every day. Suitable pasture areas were necessary, so adequate fencing had to be provided. Corn was used to supplement the grazing most of the year, since the pasture areas were generally not of the highest quality.

Periodically, we would "grow out" a couple of hogs for slaughter. Daily feeding was required with corn being the primary food but "slop" from table scraps, dishwater, some milk, and excess garden products served as an adequate supplement. Two or three times a year there would be a butchering of the hogs or a yearling. These were all-day affairs with the necessity for immediately getting the freshly cut meat hung in the smokehouse for the smoke curing process. Since butchering tasks were large undertakings involving significant time and heavy lifting, they were sort of community events with a few of the neighbors helping in return for a portion of the meat and a similar return of help. This swapping of work was typically measured in units of one-half to full days, an exchange of time as workloads came about.

A flock of twenty to twenty-five chickens was always a key component of our farm life, providing an ongoing source of eggs and meat. They were free-range chickens, but by providing them with a safe site to roost and by feeding them shelled corn each evening around the barn area, they always ranged close by, scratching through the leaves and grass to find seeds, insects, and almost any other organic material to eat. Each day at the time of milking and other late afternoon chores, we would "shuck and shell" some corn for the chickens. While calling, "chick, chick, chicckk-oooo," we'd toss the corn around the barnyard in a wide sweeping motion, bringing the chickens running. Soon after, as dusk would begin, they'd all start moving to the chicken house and "fly up" on the roosting site, which consisted of a sequence of cross poles, each about a foot or so higher and further back than the one in front of it to create a design similar to stadium benches. Having the roosting site in our backyard helped to protect the chickens from predators. So did the dogs we typically kept in our yard. If a dog ever acquired a taste for chicken or eggs, however, that dog would be ruined for the chickens and never again allowed within close proximity of them.

Nesting sites—box-like structures with an open side and with some straw on the bottom—for the hens to lay their eggs were spaced around the barn and house areas, ideally on a side wall, up a bit but reachable. This placement reduced the access of snakes and other animals that desired the eggs. The hens would adapt to sit in these nests for a while during the day for the purpose of laying their eggs. Some hens would lay an egg a day; others less often. When a hen would lay an egg, she'd leave the nest making the typical cackling sound, which we'd hear from time to time throughout the day. Toward the end of each day one of the chores would be to collect the eggs from the nests.

Naturally, with fried chicken being one of the hallmark dishes of the south, one or more of the chickens were regularly selected to be processed as dinner. Chicken and dumplings was another favorite. These dinners were prepared especially when we had company or when there was a special occasion, like a dinner-on-the-ground church service. Since the chickens fed on such a wide variety of things, Mother was very adamant about the targeted chicken being placed in a holding coop for several days where it could be fed only corn and other clean food to purge its system of the dirty food. The standard method for taking the chicken's life seems very brutal today, but was the common practice then. The chicken was quickly decapitated by holding the head in one hand and quickly tossing the body in a circular motion. This is termed "wringing the chicken's neck" and I remember being instructed many times to "go wring the chicken's neck." It was an unpleasant task, but a necessary one. Though such practices, in one form or the other, occur today on a daily, widespread basis, they take place in the backside of slaughterhouses. We're sheltered from this component of the process, consuming chicken today in blissful ignorance.

We could afford to periodically sacrifice chickens from the flock because they were quite prolific in replenishing themselves. Routinely, certain of the hens would go into the brooding or "setting" mode where they would lay a clutch of ten to twelve eggs in a nest and stay on the nest continuously for the purpose of hatching chicks, about a three-week incubation period. Since we had several roosters in the flock, the eggs were almost always fertilized and would hatch. I learned early how the term "ill as an ol' setting hen" came about: a setting hen would aggressively protect her clutch of chicks.

Today, there's a trend in the restaurant industry called "farm to table." We knew nothing else. In my childhood, our farming practices were not designed to supplement our food supply with the occasional farm-fresh meal. Our farming practices represented the source of our entire food supply, as well as our income. Very simply, it was how we survived. It would have been easy, then, to get lost in the work, to focus completely on the day-to-day effort needed to survive and to forget what a miracle it really was that nature was able to provide us with what we needed, so long as we paid attention to her. All of us were grateful for what we could produce, of course, but in addition to gratitude, I found myself, for whatever reason, feeling something more. I carried with me, from as early an age as I can remember, a constant curiosity about how it all really worked. I didn't know any other ways but nature's ways in those early years, and so it would have been easy to take nature for granted. But there was an inquisitiveness inside of me that became something of an obsession. As much as I knew and had learned about the natural world that I was a part of, I badly needed to know more.

CHAPTER 4

Discovering the Birds and the Bees

A "pecking order" isn't just an expression that describes the hierarchy of some particular group. In fact, as a young boy on our farm, I didn't know it was an expression at all. It was a real thing and the word "peck" meant something very specific. The expression is sourced in the particular hierarchal structure that I witnessed firsthand in the yard where the chickens scratched about. The pecking order among our flock of chickens was set rather simply: if one particular bird could peck at another without fear of retribution, that bird was higher on the pecking order. Only later in life would I hear people using the phrase metaphorically and in other contexts.

Among my early discoveries of the ways in which nature works was one I was introduced to by my granddaddy, Wallace Homer Carruth (Wallace is where the "W" comes from in my name). "Big Daddy," as we grandchildren called him, was a retired welder from the Illinois Central Railroad, a smart man and a good craftsman, proficient at sketching and building things. Big Daddy taught me something fascinating about the peck order of our chickens with a fun trick we'd play on the roosters.

What I learned is that if you take the top rooster in the pecking order and camouflage his neck feathers, he's no longer at the top. I had hours of fun with this. I'd take the respective rooster to the wash pot, take a moist rag and rub it on the pot to dirty it with soot, or smut as it's often called, then rub that smut on the rooster's neck. I'd let him return to the yard where we always had four or five other roosters around and he'd go strutting in as always, only to discover to his chagrin that he was no longer the de facto leader. Whether the smutting meant that he somehow went unrecognized by the others is something I still don't know for sure, but it sure seemed as if that rooster was now a stranger to them. And they took umbrage at a stranger coming in trying to upset the apple cart, as it were. The now-dethroned rooster would get chased and pecked and have to start his climb back to the top all over again.

A pecking order, which is common throughout the animal kingdom (and in the human realm, for that matter), is really

Placing smut from the wash pot on the neck feathers of a rooster.

nature's way of avoiding chaos. Without it, our roosters would have continually been all over each other in their efforts to service the hens about the yard. With the order came territoriality and organization. The barnyard with the bulk of the hens was the top rooster's domain. Second in order might find himself on the outskirts of the yard with a much smaller share of hens; third might find himself closer to the pasture with only a couple of hens. But each knew where he belonged. Very little came along to disturb this order and life for the hens and roosters went smoothly along.

Until, that is, I would mischievously smut the leader's neck. Then it was anarchy, at least until the order could sort itself out once again. It was my own little way of messing with nature, intervening in a natural process, and learning the results of doing so, namely, chaos amongst the chickens. In the end, more often than not, the leader would regain his position after a few decisive fights. Sometimes he'd be unseated, however, and not so much strut out of the yard as slink, his pride wounded more than his body. These were never fights to the death, or even serious injury. There was a lot of posturing and crowing and flapping of wings. One would always surrender before the fight got out of hand. The cockfights most people think of, with a crowd of people looking on and making bets, is a decidedly unnatural situation, typically brought about by keeping the roosters in impossibly tight quarters, pumping them with steroids, and breeding and training them specifically to fight.

For a few years, I kept a rooster as a pet. A brightly colored baby chick that Patsy and I obtained for Easter one year turned out to be a male, and that rooster followed me around. Patsy and I eventually decided to introduce him into the yard with the others and he managed to find his place, although he was quite a bit tamer than the rest. I still don't know how it happened, but one day I came out to the yard and noticed my rooster had somehow become the boss rooster. It was enough to make a fellow proud.

Big Daddy Carruth also taught me how to trap birds, typically ground-dwelling birds like thrashers, by building a wooden cage-like box turned upside down. You'd prop one end up with a triggering device consisting of three sticks appropriately carved and assembled in a certain way that would cause the box to come down over the bird once the bird pecked away at a piece of bait, maybe a hunk of bread, that you'd affix to the trigger. Big Daddy admonished me to handle the birds gently and to release them promptly after a brief up-close examination. I always enjoyed watching them scoot away, quickly, and surely a bit more cautious about where they dined, at least for a few days.

Big Daddy also showed me how to call turkeys and hawks. For hawks, you'd take a pencil-sized twig, cut a slice in one end and insert a piece of cellophane, which you'd hold in place with a rubber band. By blowing on the side of it, that cellophane would vibrate along that slit and produce a sound just like the call of a hawk during mating season. For turkeys, you'd select a leaf of the proper texture, and by blowing on a folded-over portion held to your mouth in just the right way, you could mimic the characteristic "yelp" and "putt" calls of a hen turkey, which can elicit the "gobble gobble" answer of a male during mating season in the spring. Hunters use these calls to lure the aroused wild gobblers during hunting season. Today, you can buy a number of commercial makes and styles of turkey callers at any hunting goods store or even your local Walmart, some of them electronic. I'm still somewhat partial to Big Daddy's method.

Around the property, we always had martins (more specifically called purple martins) and not because we called them, but because Daddy would erect a specially designed house with several small compartments on a pole in our yard. Sometimes he'd use a cluster of gourds. By providing these housing arrangements, we could expect a group of martins to arrive each spring, raise a family of youngsters through the summer, and stay until fall, when they would depart for their winter home.[2] The martins, small as they were, were especially aggressive and kept the chickens safe from predatory hawks. It never seemed to matter how big the hawk or how small the martin. It was another natural solution to a natural problem. Many times, I'd see a hawk swoop close by and then see him swoop right out again, chased by several birds a fifth his size in what looked like some kind of World War I dogfight, the large hawk making evasive maneuvers and the little martins close on his tail.

There was a particular kind of bird we knew as a bee martin. Actually called eastern kingbirds, these birds would feed on bees and wasps. We also had a type of bird we called the French mockingbird (which I would later learn was also called a shrike) that would catch insects and small animals like lizards and impale them on our barbed wire fences so they could get at the bodies easier or else leave them for a later meal. We never cared for the French mockingbirds because they would attempt to prey, sometimes successfully, on the young of other birds, even the martins.

In addition to keeping martins around to protect the chickens, Daddy liked keeping a snake around to protect the corn. A corn snake could be a useful employee. Corn snakes feed on mice and rats and if Daddy came across one, he'd grab it and put it in the corn crib. Mother went out to feed the chickens one time and grabbed an ear of corn from the crib only to be startled by the presence of a snake she was not expecting. "Ferd!" she hollered, storming out of the crib, "why didn't you tell me you put a snake in there?!" It wasn't the snake that

---

2   Purple martins are a type of swallow. They overwinter in Brazil and return to their North American habitats in the spring, and will return to their previous successful nesting locations. Purple martins themselves can actually be targeted prey for hawks and owls, so placement of their houses in an open area away from trees is important so that they have space for defensive maneuvers.

bothered her; it was not being told about it beforehand.

Our natural setting provided unending discoveries that kept me fascinated, many of them passed down to me by my elders, like Big Daddy Carruth with birds. Or Uncle Farrar, my mother's youngest brother, with bees. Uncle Farrar was a big man that I looked up to in more ways than one. He joined the Army after high school and went off to Korea, though, fortunately, he was never in direct military conflict. Before he went, he taught me how to catch bees without getting stung. The best way? Catch a bee that doesn't have a stinger.

As it turns out, male carpenter bees—we knew them as whitehead bumblebees—are stinger-less, the stinger being an adaptation of a female bee's reproductive system. You could spot a male carpenter bee, Uncle Farrar taught me, by looking for the characteristic white spot on his head. These bees could often be found hovering around the eaves of the barn or the house and if you were quick, you could nab them from the air and hold them in the palm of your hand. I made more than a few mistakes in identification, sometimes snatching a female carpenter bee, errantly thinking I'd seen the distinctive white spot. Female bees are always quick to let you know of a mistake of this kind.

Once you had a male bee in the palm of your hand, the thing to do would be to show it off to your friends by putting your palm up against their ears so they could hear the buzzing and be impressed that you had the nerve to hold a bee in your hand. If that didn't impress, you could go a step further. Uncle Farrar showed me how you could take a length of thread, fashion one end into a tiny lasso, and gently slip it—not too tightly—around the bee between its abdomen and thorax. Then you could just release the bee and amaze everybody as it hovered in the air, like a dog on a leash, at the end of the thread you were holding. You'd either amaze or frighten your friends. It was always great fun to chase the other kids with the tethered,

Watching the hovering flight of a male carpenter bee tethered on end of a thread—soon released to zoom back to the wild, of course.

buzzing bee. Of course, after a few laughs, it was important to untie the bee and let it zoom back into the wild.

My father, of course, had taught me how to catch doodle-bugs and dragonflies, but he taught me about trapping bigger things, too. To help supplement our income, Daddy would set traps along the Amite River for mink. Back in those days, you could send a mink pelt to Sears & Roebuck and they'd grade it, and then send you wholesale payment for it. That's where the mink stoles came from for their stores and catalogue. There was something of an art to trapping and I would follow Daddy along the river and watch with fascination as he'd set a trap and then throw just the right amount of mud and water around it to cover our tracks and scent.

With all of nature's creatures, whether it was mink, roosters, bees or whatever else I was learning about, what interested me the most were the behaviors. The turkey call was fun to produce, but the fascinating thing was how male gobblers would respond back to the right sounds of hen turkeys, and how if you hid properly in the right locations in the woods, you could actually lure otherwise very sly, wild gobblers right to you. With a lassoed bee, I studied up close the aerodynamics of how the bees buzzed around. Nature was filled with treasures, a never-ending supply of them, it seemed to me. The closer I studied these treasures, the more interesting they became.

Etched forever in my memory is the first time I observed a bat up close and personal. I was six. One evening after dark, my parents and Patsy and I were at Uncle Lonnie's grocery store. While my cousin Paul and I were inside and the others were on the porch, a bat flew into the store. Paul and I began swatting at it with brooms, eventually knocking it to the floor. When I pounced down and grabbed it, I was stunned by a piercing bite between my thumb and index finger. To this day I feel fortunate to have somehow avoided rabies. I saw for the first time that this creature was not like the other flying creatures I had seen in my six years. This was no bird. This had the face of a rodent, with hair and ears and teeth. Sharp teeth, as a matter of fact.

One reason I might have been surprised is that the actions of the bats I had seen to that time had been mimicked by a bird known as a nighthawk. The nighthawk is a bird that feeds at night and flies erratically, like a bat, making long dives and just as quickly putting on the brakes and pulling up and darting in another direction, its wings making bull-like sounds. We knew them as bull-bats, in fact, so it's easy to see what I had in my mind when I first approached the bat that night in Uncle Lonnie's store. But the creature I encountered had neither feathers nor a beak.

What it did have, as I would learn, was something called radar. Until then, I assumed that, like a bird of prey, a bat would visually spot its target. We often used to throw little pebbles up the air at dusk to watch bats dive for them, only to veer off once they realized the pebbles were not insects. It was stunning to learn that a bat actually picked up the motion of the pebble from an internal radar system. Nature just kept getting more and more interesting.

I kept learning about insects, too. There was another kind of doodlebug I learned to catch, another member of what I would one day understand was part of the *Neuroptera* insect order. This doodlebug was called an antlion, part of the family *Myrmeleontidae* of the order *Neuroptera* (big names I would learn much more about, much later in life). In the larval stage, the antlion has a plump little body with long mandibles. In sandy soil, they build little pits that trap passing ants or other prey. The antlion feels the vibration of the ant sliding into the pit and is ready to grab it with his mandibles. If you know what you're looking for, you can spot these little cone-shaped traps, tap on the side of one with a piece of straw to mimic a fallen, trapped ant and snatch the antlion right out of the pit just as he closes his mandibles on the straw.

I also began to observe and learn about different types of "hunting wasps" that catch various insects and spiders and take them back to their nests for their larvae to feed on. These wasps included dirt daubers, a type of wasp that builds its nests out of mud; the paper wasps that build their gray paper-like nests in sheltered places such as under the eaves of houses, the underside of tree branches, or the open end of pipes; and the yellowjackets that also build similar papery nests in protected structures like tree stumps and cavities in the ground. On numerous occasions, due to my unrelenting curiosity, I discovered in a very real way that the paper wasps and yellowjackets can be quite aggressive if disturbed, and both pack a painful sting. Later I would learn that the dirt daubers are classed in the family *Sphecidae* and the paper wasps and yellowjackets are in the family *Vespidae*. All of the wasps and bees are part of the order *Hymenoptera* (other big words I would eventually learn).

Of course there were the lightning bugs, perhaps the insect that intrigued me the earliest in life because of its natural magic lamp. I would go on to learn how efficient these insects are, generating light while losing very little energy to heat, something we humans could never do with our own incandescent lighting and something we're only now beginning to approach with LED lighting. Lightning bugs are still a subject of study as nature's unmatched model, guiding pursuits of even more efficient energy usage.

Larger wildlife that grabbed my interest included foxes. Daddy hunted foxes and we always had fox-hunting dogs, along with dogs for hunting quail and squirrels. The fox hunting was mostly a social event with cousins and other people from the community, everyone more enamored by the chase than anything else, and mostly content to sit around an evening bonfire listening to the dogs. The men all knew their dogs by the sounds of their barks and could tell which one, having come across the scent of a fox, might be yelping, which ones were participating, and which one was leading the chase. Once a dog got on the trail, there was no stopping him.

I saw this kind of stubborn determination with our dog Bobo when I'd go out hunting squirrels with him. He'd find the scent and chase a squirrel up a tree and stay there, looking upwards and barking until he knew you understood exactly what it was he'd found for you. Sometimes a squirrel would be smart enough to leap to the branch of another tree without Bobo noticing and make its escape. For their part, the foxes had their own tricks when being chased through the woods. They'd often run through water to mask their scent or sometimes loop around just to confuse the dog.

Squirrel, quail, or fox, it was the hunted and the hunter, perhaps the most basic of all natural relationships, next to mating. What I learned with all of my observing and discovering was that when it came to hunting, there were really only two ways for a creature to proceed: ambush or search and find. The antlion uses the ambush technique. Dirt daubers use search and find. As humans, we copy these animal techniques, building a bird or mink trap or using a turkey call as an ambush, or setting the dogs free to search and find a fox. We've never done anything that nature didn't do first. Maybe this is what interested me in nature in the first place, even at such a young age. Nature was infinitely wise and I couldn't help but feel overwhelmed with admiration and respect. Everything was so efficient and every animal so resourceful. My insatiable thirst for learning about the environment around me continued and, as I grew older, I got it in my head, somewhere along the line, that no matter what I was going to do or be as a grownup, it was going to have to include continued study of the natural treasures of my world.

CHAPTER 5

A Time of Change

In 1940, the year my parents married and began forming our family, there were over six million farms in the United States. By the turn of the century, in 2000, there were around two million.[3] The biggest drop came during the 1950s through the 1970s. These were the post-war boom years. Mechanization increased by leaps and bounds and farms expanded in size while consolidating to take advantage of efficiencies of scale. The accounting rules were changed to allow capitalization of large purchases, making it easier to replace horses and mules. The age of specialization, centralization, and big machinery had arrived.

Although there was a huge drop-off in the number of farms, roughly the same amount of land was still being farmed. By necessity, the individual farm had to become larger and larger to compete. The average farm size increased from 175 acres in 1940 to over 400 acres by the year 2000, and even more telling is the fact that by 1993, almost half of gross farm sales came from farms with a mean acreage near 3,000.[4] This trend toward

---

3   Data from History of American Agriculture 1776-1990, USDA, Economic Research Service, Washington, DC, POST 11, and USDA.

4   Robert A. Hoppe, Robert Green, David Banker, Judith Z. Kalbacher, and Susan E. Bently, *Structural and Financial Characteristics of U.S. Farms, 1993*, Department of Agriculture, Economic Research Service, Agriculture Information Bulletin, no. 728, p. 30.

gigantism was a phenomenon that was happening in every industry. Earl Butz, secretary of agriculture under the Nixon administration, admonished farmers to plant "fencerow to fencerow," and put it in even more blunt terms: "Get big or get out."

It's not hard to imagine where that put our little farm. The sharecropping days were numbered. Having just five acres to grow your main cash crop was untenable. Keeping chickens for eggs, milking your cows, growing vegetables for your own family table—well, that all quickly became the quaint idea of a bygone era. People no longer got their milk from the family cow. You went to the supermarket and bought a gallon that might have come from a tank that held the milk of a hundred cows. You bought shrink-wrapped meat that might have traveled two-thousand miles or more. You bought chicken that came from multi-million-dollar Perdue processing plants. Ole Nell, our milk cows, the flock of chickens, the roosters and their peck order that had so fascinated me—our farm, our way of life, had no place in this new world.

By the mid-1950s when this shift was reaching full swing, I was a teenager with my whole life ahead of me and, therefore, numerous options. My father, on the other hand, was in his late forties, a sharecropper his whole adult life who was unable to read or write. He'd always hid his illiteracy well from most people—maybe saying he'd forgotten his glasses if someone put something in front of him to read, or counting on Mother to cover for him—but there was no hiding it during a job search. To watch up close and personal the decline of the family farm, and thus my Daddy's ability to take care of his own family, was painful and heartbreaking. Daddy would hang on as long as he could, but by 1957, his last year of farming, the income he brought in for the year was a grand total of forty-eight dollars.

We all saw the writing on the wall, even if we didn't quite understand the global shift as it was happening. We shared in some of the technological progress of the time—we finally moved into a house with electric lights and even obtained a small propane cooking stove, though we still had no running water—but for the most part, the times were leaving us behind. Daddy tried to find work where he could to supplement the farm income, traveling around south Mississippi and even going down to Louisiana. People were finding factory or trade work in places like Natchez and Baton Rouge, but Daddy came home empty-handed, unable to latch onto anything for which he was qualified. Mother went to work as a seamstress in a garment factory in McComb, carpooling with other women from our area, and making seventy-five cents an hour. She maintained her household duties, too—tending the garden, cooking, and washing. It was important to her to send us off to school in clean clothes, cleanliness being "next to godliness." I had two pairs of jeans, which meant Mother was typically washing one pair every evening, although some days, if I managed to keep my jeans especially neat, she'd let them go one more day. She'd use the stove to clean them, heating up water that came not from a spigot, but water that she had drawn from the well. For larger loads, she'd fire up that cast-iron pot. Then she'd starch and iron the jeans for a smart crease. Shirts, too.

My parents tried to keep the household financial situation from Patsy and me, but we knew what was happening. We'd get twenty-five cents a week allowance and probably could have asked for more if we were in some kind of bind, but we never asked. We knew the hardships we were facing as a family and I stretched that twenty-five cents as far as I could. There was a little store across from my school that sold ice cream. Five cents for one scoop, ten cents for two. I'd agonize over the decision; the two scoops always looked mighty good. In the end, I'd typically settle for the single scoop and keep the other nickel for another day.

When I was a senior in high school, Daddy finally got a job. We both did. The school needed bus drivers and we became a team, of sorts. As the main driver, Daddy made ninety dollars a month and as his substitute driver, I made forty-five. Before being hired, however, we both had to pass a test. A written test. We'd applied for the positions and were approved, but then we needed to attend a one-day program where we were made familiar with the rules and regulations. There was a driving portion of the test that we both passed with no problems, but then came that written portion. I sat with Daddy, reading him the questions as he checked off the answers. The instructor watched us and, in an act of wonderful generosity that I remain grateful for to this day, said not a word. He understood. He knew what failing my illiterate father, or even intervening, would have done to his pride.

I ended up driving the bus frequently, since I was going that way, anyway, and getting some karmic payback for the times I'd misbehaved on the bus myself getting in a fight here or there or launching the occasional spitball. Now it was all happening behind my back and I had to try to drive and maintain discipline at the same time. All in a day's work, I supposed. In the summer before my senior year of school, I took another job, knowing I needed to find a way to help the household even more. I was getting older and my needs—clothing, spending money—were increasing just at the time when my parents' means were at their lowest. I knew there weren't a lot of opportunities where we were, so I headed for Natchez, where my father had looked, and found a job in the Johns Manville plant at $1.85 per hour, more money than my mother or father had ever made.

Academically, I did well in high school, a continuation of the success I'd had since grade school. Things always came easily to me. Good genes, I imagine. My father may have been illiterate, but he was smart, and Mother even more so. The only thing I ever struggled with was penmanship. I could never understand why, but I couldn't seem to write as neatly as the other kids. Years later, I would see my first left-handed desk and it all made immediate sense. For all those years, I'd been writing with my left hand using right-handed desks, the only kind of desk the school had available.

Perhaps more important than my genes, I was blessed with good teachers. They weren't highly technically trained, but our teachers understood their students. They knew us, they knew the area, they knew the families we came from. They understood the culture and they taught within that context. They understood well Maslow's hierarchy, the imperative of having basic needs met before such luxuries as learning. They got to know the kids. There weren't many of us. During my senior year, there were 311 students in the first through twelfth grades and only seventeen in my class. The teachers knew who was struggling or who was being bullied or who was having problems at home. They understood that the context was as important as the content. They knew how to teach *us*. Mrs. Gill, my English teacher, for instance (the only English teacher at West Lincoln for many years), didn't know how to diagram a sentence, but she taught us another way to identify the parts of a sentence and we learned as well as anyone possibly could have. Later, in a local junior college English class, the instructor would say that the West Lincoln graduates didn't have to bother with diagramming sentences; she was aware that we had been taught another way that worked just fine. In other words, my high school education was customized, built on a culture of strong, hands-on relations. This approach ran all the way through the local educational system, on up to the board of trustees who were community members first and foremost. Our educations were the better for it.

Of course, "hands-on" didn't always mean in a pleasant way. Our principal kept a big razor strap in his office and discipline was sure and swift. There was a sort of informal peer system, too, where we students would more or less police ourselves, the upper-classmen not shy about taking a rogue kid out behind the gym to teach him the way things worked. It's hard to say which induced more fear, the razor strap or the trip behind the gym, but both were effective means of keeping order on a very local basis. I never saw the sheriff or any police officer ever called to the school.

I did well in Mrs. Gill's English class where we frequently wrote essays. Later, in college, a professor would tell me I expressed myself well with words, which I attribute to Mrs. Gill. We had civics class where we learned about the federal government, but we also learned about local politics and maybe it was here where my interest in community involvement later in life was first stoked. One day, Mr. Herring talked to us about corruption, pointing to an earlier period of time when it was pretty well known that all of our county supervisors had taken kickbacks from big equipment companies. "There was only one who didn't," he said. "And that was G.W. Lewis." G.W. was that great uncle of mine who preached at Mount Gilead. "Joe," Mr. Herring said, looking at me and causing the class to turn my way as well, "your Uncle G.W. never sold out to anyone. He told them 'that's the people's money, so we'll buy good machinery at a fair price and let the people keep any money that's not needed.' You should be proud." Of course, I was, and that heritage from my Uncle G.W. would serve as a guiding compass for me in entrusted community roles I would take in later years.

In addition to English and civics, there was math and science. I liked biology class the best, the place where we studied those miracles of nature that I had been observing all of my young life.

Our school was too small for a football team, but we had baseball and basketball and I did well in both. I was class president as well as valedictorian in my senior year, and it started to look more and more as if I was heading for a life of academia. Certainly, I was headed for college. My parents had made that clear. Daddy had learned it the hard way. "They can never take your education away from you," he would tell me, witnessing how his own way of life was being stripped from him a little more each day.

Also nudging me towards college were C.V. Linton, the principal and math teacher, and Wyatt Tullos, my biology teacher and baseball coach. Both apparently saw potential and took an interest in my future. Coach Tullos especially encouraged my interest in biology. "You need to be in life sciences," he would tell me. "They need you in that field." Both talked to me about scholarships. "You're a good hitter, and maybe the best catcher I've ever coached, so we can get you a baseball scholarship, if needed," the coach said, "but your real strength is in the academics, so I want you to understand, you are a student first, athlete second."

I was the baseball team's catcher, senior year in high school.

Nevertheless, the idea of a science *profession*, to pursue the subject as a career field, seemed distant and abstract to me. I couldn't quite relate to the notion. But I was inspired and motivated by the idea that Coach Tullos and Principal Linton believed in me. These men were role models and their guidance meant a lot to me. Years later, I would read a quote by Colin Powell: "Enriched by our diversity, strengthened by our trials, inspired by our dreams, we are a nation where generation after generation of ordinary people performs extraordinary acts of accomplishment." It reminded me of just where these "ordinary" people come from. They come inspired by people like Tullos and Linton and an accessible school system that can provide opportunity for all, regardless of geography or wealth or social standing. Our nation's public education system might just be our most democratic institution. And a necessary part of anybody's definition of the American dream. I would have been nothing without it. I would have gone nowhere. My teachers at West Lincoln are all gone now, but whenever I'm called upon to speak to a group of teachers, I make sure to tell them how I feel about the people in their profession: they are my heroes.

As I graduated high school, heading for junior college, my father was still seeking work. As a shot in the dark, he had Mother write a letter to John D. Smith, the south Mississippi Highway Commissioner, a man Daddy had never met. The letter explained Daddy's situation and made a plea for a job to help his family make ends meet. Somehow this letter caught the commissioner's eye. One weekend, a car came driving up to our house. A man from the district office of the highway department in McComb stepped out and told my father, "The commissioner sent me to see you. He instructed me to tell you to come to work Monday. You have a job."

Daddy started in the mechanic's shop, helping to grease vehicles, change oil, and take care of other maintenance issues. He would work there, earning a living wage, for long enough to qualify for a retirement pension, always careful to live up to the commissioner's goodwill. Within a couple of years, Mother and Daddy were able to purchase a modest house along with five acres of land where they lived comfortably for the balance of their lives. To this day, I remain profoundly grateful to John D. Smith for his simple outreach of kindness and the immense benefit it provided to my family. I try always to keep that outreach in mind as opportunities to touch others come my way.

With the kind help of a stranger, my father found a way to get by, even as the life we'd known disappeared before us. The world was changing in irreversible ways. Urbanization was taking hold as the small family farms died out, changing the dynamics not just of farming, but of the family unit itself. People didn't go off to the fields adjacent to their homes in the morning; they commuted to their jobs across town, or in another city. Everything became more specialized, too. Farms began focusing their output on single crops, and people became specialists, as well. You had a single job to perform at places like Johns Manville. The country doctor I knew would eventually be replaced by people with particular specialties—internists, dermatologists, ENTs, cardiologists, obstetricians. House calls became a thing of the past. It was a time that began to see the great growth of centralization. The company you went to work for was probably headquartered in another part of the country. The local general store was now a branch of a large chain. The local school slid away from the customized "hands-on" approach that had benefited me so well, and would eventually become guided by state and national standards. Globalization was on its way and, ostensibly, greater interconnection with larger amounts of people. It was a time of great progress. Much was about to be gained by the world as we entered the second half of the twentieth century. I couldn't see it at the time, but later I would come to reflect on those early days of change and realize that much was also about to be lost.

CHAPTER 6

A New World

On a warm breezy day in early September of 1962, I arrived on the campus of Mississippi State University for the first time. I had completed my two-year associate's degree in junior college and enrolled in MSU with plans to continue pursuing my interests in life sciences. I soon settled into my classes, finding the university very much to my liking. I enjoyed getting to know the faculty and my fellow students and I loved the rolling hills of the campus, especially as the fall weather and colors began to appear. Little did I know what this charming setting had in store for me—the opportunities and experiences that would soon come my way and continue to come my way at an ever-increasing rate over the next five years, and the pivotal, lasting role they would play in my life.

I spent the first semester completing general coursework, but made it a point to use this time to visit the various life sciences-related departments. I wanted to explore the types and availability of careers associated with zoology, botany, genetics, microbiology, immunology, wildlife management—they all had some appeal to me. After a few discussions, however, entomology—the study of insects—which had interested me since those days when I had chanted, "Doodle, doodle, your house is on fire!," began to stand out with special appeal.

One of the departmental people I talked to was a gentleman by the name of David Young, an extension service leader in entomology. Interestingly, I would discover during our discussions that David was the brother of L.G. Young, my vocational agricultural teacher in high school. Small world. In talking with David, along with Arlie Wilson, another faculty member in entomology, I learned of several positive developments for entomology at Mississippi State, some recent and others that were in the works. It so happened that a newly structured entomology department had just been formed, and a professor by the name of J.R. Brazzel had been recruited to serve as its head. Brazzel would be arriving on campus within the next few weeks. Previously, entomology had been a joint department, with zoology, as part of the College of Arts and Sciences. Now, it was being split off into its own department, to become part of the College of Agriculture. Additionally, a modern, state-of-the-art boll weevil research laboratory had just been built on the campus by the USDA, Agricultural Research Service. This new laboratory was being staffed by a substantial array of scientists, several of whom would also serve as faculty for the new department. Clearly, great things were in store for entomology at MSU.

With notable timing, these developments were playing themselves out in the midst of what would be a watershed moment in environmental science. In September of that very year, 1962, Rachel Carson published her seminal work *Silent Spring*. The book documented the harmful effects caused by the expanding use, and often arbitrary overuse, of recently discovered, broad-spectrum synthetic pesticides for crops, such as DDT. The book inspired the environmental movement and would ultimately help lead to the creation of the United States Environmental Protection Agency. Earlier that year, *The New Yorker* had published the book in serial form and Houghton Mifflin, the publisher, had released proofs to many, including delegates of the White House Conference on Conservation, so Carson's work was already a major topic of conversation by the time the book hit the shelves in September and I hit the campus of MSU.

There was quite a bit of controversy over *Silent Spring*, with all the major chemical companies vehemently criticizing Carson's research. Regardless of one's position on the matter, the debate was spurring a growing awareness of the potential hazards associated with indiscriminate use of pesticides, and the need for a more balanced approach built on a better scientific foundation. In a larger sense, *Silent Spring* represented nothing less than the start of a national conversation about the environment—humankind's place in it and level of responsibility for it. It was with this as a backdrop that I was setting my sights on a career in entomology.

Dr. J. R. Brazzel, entomology department head, and major professor during master's work. I will never forget the passion he instilled and the pivotal influence of our first meeting

I met with Dr. Brazzel, the new department head, shortly after he arrived on campus. He was vigorous and enthusiastic and had become well known in academic circles. He was the perfect person to come along in the new "Silent Spring" era, and it was quite a coup for MSU to land him. If there had been any doubt about my direction, it was gone by the end of our visit. I felt Dr. Brazzel's fervor and energy the moment I walked into his office. I remember that he had a cigar in his mouth that curiously kept getting shorter despite the fact that he never lit it. But I was riveted by our conversation. We talked for quite a while. It's been said of certain people that you may not remember exactly what they said, but you will never forget how they made you feel. I do still remember much of what Dr. Brazzel said, but far more importantly, I will never forget how he made me feel that day. He asked about my background and my interests and studied my resume and my school records. He shared with me his visions for the department and how they related to the field and directions of entomology. Then, after inquiring a bit more about my goals and ambitions, he paused and stated to me emphatically, "Joe, you have the right stuff, and we want you in our program. If you choose to join us, we will support and prepare you for a bright future in this field." I remember the passion with which he spoke about entomology and agriculture and the strong sense of belonging that I began to feel. It was as though Dr. Brazzel reached out and planted a portion of his enthusiasm into the center of my psyche.

Dr. Brazzel would make good on his promise. Not long ago, I was presented the Distinguished Fellow Award by the MSU College of Agriculture and Life Sciences, the purpose of which is to recognize alumni who have exhibited significant professional accomplishments and demonstrated the qualities and traits that the university endeavors to instill in its students. I had come full circle and remembered clearly Dr. Brazzel's words to me back in 1962.

I left Dr. Brazzel's office that magical day feeling energized and focused, and entomology not only became my major area of study from that point on, I decided to take my pursuit all the way through to a Ph.D. Immediately, I started implementing a game plan to that end. I earned my bachelor's degree in June of 1964, being awarded the "Senior of the Year Award," and directly began work on my master's, which I finished a year later. Then it was on to my Ph.D., which I completed in September of 1967.

Dr. Brazzel was my major professor during my master's work and Dr. Bradleigh Vinson was my professor during my

Dr. Bradleigh Vinson, major professor during doctorial work, an innovative thinker who taught me to take bold and creative approaches

doctorate work. Bradleigh and I had been students together before he'd acquired his own Ph.D. at MSU and joined the faculty. We had jointly written and published a research paper while I'd been getting my master's and we'd developed a synergy that made it very rewarding to work together. I ended up being Bradleigh's first Ph.D. student. He was an innovative thinker and taught me to take bold and creative approaches. He had considerably more experience than I did and he shared it freely with me, becoming something of a big-brother mentor. Our close friendship and collaboration continued for years, even to this day.

Starting in my junior year, right after my visit with Dr. Brazzel, I was provided a job working with the entomology department on insect-rearing and other laboratory work. The department also provided me an office and lab space with the graduate students so I was able to be a part of this group and to start my research for my master's even before I graduated with my bachelor's. This environment was immensely stimulating. We held many vigorous, challenging conversations, sometimes well into the night, on a wide range of professionally and non-professionally related subjects including science, agriculture, environmental issues, politics, religion, sports, current events, historical and modern trends—nothing was outside of our purview. Numerous viewpoints were represented as we were of widely diverse backgrounds. I could sense my horizons expanding and my intellectual grasp of varying subjects rapidly growing.

Once a graduate, I was provided a full graduate assistant-ship, working with the department's research and teaching programs and receiving funds to cover school and living expenses. These undergraduate and graduate days at Mississippi State University were exciting times for me. Besides the widening of my horizons, it was a dream come true to be able to study the creatures that had so intrigued me as a child in comprehensive, academic ways in the classroom, field, and in the laboratory. In my more general classes, I first learned about the physiology, genetics, reproduction, behavior, anatomy, classification methods, distributions, and more of all sorts of animals and plants. The studies included vertebrates, invertebrates, birds, fish, amphibians, reptiles, mammals, marsupials, and everything else.

As I began to specialize in entomology, I studied insects in much greater depth and learned a great deal about their immense significance to humankind in both beneficial and harmful ways. Of all the different species of animals known to exist in the world, three quarters of them are insects. I would learn about the major classification orders of insects including *Coleoptera* (beetles), *Lepidotera* (butterflies and moths), *Orthoptera* (grasshoppers and crickets), *Hymenoptera* (bees and wasps), *Diptera* (flies), *Odonata* (dragonflies and damselflies), *Hemiptera* (true bugs), *Homoptera* (aphids, cicadas), and numerous other orders and other aspects of the system of classification for these insects and how to recognize them.

I learned about the highly diverse and numerous species that are adapted for almost every kind of habitat, from the North Pole to the tropics. Some species possess antifreeze type material and can withstand subfreezing conditions while others are adapted for hot desert conditions. Many are harmful and compete for our food, fiber, and shelter, and are able to feed and live on all manner of plants, while others live and feed on dung, carrion, cloth, and even wood. Some, like mosquitos, are aquatic during the immature stage, but airborne as adults.

Some transmit serious diseases, from malaria to yellow fever. On the other hand, I learned that many types of insects are helpful and vital for our survival, such as plant pollinators, honeybees, silkworms, and the natural enemies of pest insects, such as the lady beetles.

I learned all about the bodily structure of insects. Unlike us, with our skeletons on the inside, insects have an exoskeleton, worn on the outside. Instead of using lungs, they breathe through a network of tubes, called trachea, which penetrates their bodies. Where we have closed circulatory systems, an insect's blood, called hemolymph, flows openly in its body, circulated by a tube-like structure located in the upper side of its body, called the dorsal vessel, where the hemolymph is taken in at one end and pumped out the other.

I studied the principles of insect physiology, nutrition, reproduction, and behavior. Additionally, I learned a variety of methodologies and processes for research. I learned how to collect data and do statistical analyses, how to identify patterns and real differences at specified confidence levels versus random occurrences. As part of my work, I made numerous visits to modern-day farms, observing large farming operations in the Mississippi Delta. This was my first experience of witnessing firsthand the new directions in farming, and I was struck by the stark contrast between those farms and our little family farm back home.

In addition to Brazzel and Vinson, I was the beneficiary of some wonderful professors and faculty members. Dr. Johnnie Ouzts, Dr. Jimmy Land, Dr. Leon Hepner, Arlie Wilson, Dr. Leslie Ellis, and numerous others—each playing an influential part in my education. Some would be lifelong friends of mine. My fellow classmates were special, too. I delighted in their fellowship and we fed off each other's enthusiasm. Richard Jones, Alton Walker, James H. Tumlinson, Aubrey Harris, Marcus Adair, Wendell Snow, Emery Skelton, Robert Kincade, Glenn

Worley, Howard Barefoot, Beverly Norment, Reed Dinkins, Charles Boone, Travis Pate, Clifford Hoelscher, Sammy Polles, Frank Timmons, Don Barras, Milton Ganyard, and many others, too many to name, made for a unique environment of learning and fun. Several I would work closely with later in my career and share enduring friendships.

Besides passing the necessary coursework to earn a master's or doctor of philosophy degree in science, it's required that a committee composed of faculty members approve the respective degree. For each degree, you must conduct a project of original research whereby you document the rationale for your work, your methods and materials, and findings and conclusions, all assimilated in written form to be reviewed and approved by each committee member. Your major professor (Brazzel and Vinson for me, respectively) counsels you throughout the process and coordinates the committee's activity. In the final step, you appear before the collective committee and defend the validity of your work. For a master's, this work is typically called your thesis; for a doctorate, your dissertation.

During my senior year, as a student worker in the laboratory, I was eager to get started on my thesis research work and I begin exploring the possibilities. At the time, I helped maintain laboratory colonies of certain insects for research purposes. Among those colonies were two of the most prevalent insect pests of the United States: the cotton bollworm (also, called the corn earworm) and the tobacco budworm (or simply budworm). For our rearing purposes, we used artificial laboratory-prepared food, but in the wild, the caterpillar stage of these insects feeds on a wide variety of agricultural crops and wild plants during the spring, summer, and early fall. These crops and plants include wild geranium, wild toad flax, cotton, corn, soybeans, tobacco, peanuts, tomatoes and other vegetables. Both these insects go through the egg, caterpillar, pupae, and adult moth stages. The moth deposits eggs on the plant, the eggs hatch into small caterpillars that feed on the plant, and then the caterpillars eventually go into a pupal stage in the soil, later to emerge as moths and thus continuing the cycle. The caterpillar stages of the two species look almost identical and both feed primarily on the fruit of the plant, progressively attacking larger fruit as they grow in size. Both can be highly damaging to large acres of crops.

It was already generally known that there were numerous natural enemies (diseases and other insects) that attack these two pests and help limit their numbers and damage. Natural enemies exist for any living creature. Everything has its enemies; those enemies, in turn, have enemies of their own. It's nature's way of keeping everything in balance and it's quite easy to disrupt this balance with chemical pesticides, which kill the natural enemies as well as the pest. But although it was known that the bollworm and budworm had their enemies, little was known about them, about their abundance on the various wild plants and crops, and factors affecting their numbers and effectiveness.

One of Dr. Brazzel's initiatives as department head for entomology was to increase the knowledge and awareness by the public and farming community of the importance of these natural controls, to look for ways to increase their numbers, and to encourage the cautious use of pesticides in ways that would minimize the disruption of this natural control. In keeping with this objective, I became interested in the question of which beneficial insects were important as natural enemies of the cotton bollworm and tobacco budworm, and their occurrences throughout the year.

For my master's thesis, I studied parasitic wasps in particular. These are endoparasite, creatures that live inside their hosts. The female wasp searches for a host caterpillar. When she finds one, she quickly stings and injects an egg.[5] The egg hatches into a worm stage (larva). This worm develops and grows as a "little worm" inside the "big worm," like an alien. When the larval stage of the wasps mature, they chew their way out of the caterpillars and spin cocoons from which the adult parasitic wasps emerge. They're male and female, so they'll mate, and then the females go off in search of caterpillar hosts. And on it goes.

During 1964 and the spring of 1965, we collected hundreds of caterpillars of these two pests from throughout the state. We took them from wild plants like the geranium, crimson clover, spider flower, tomato, cotton, and corn. Caterpillar collections created from fields of cotton that had been treated with pesticides were held separately from collections created from untreated fields. All caterpillars were held in the laboratory on an artificial diet to determine if they were parasitized. The parasites that developed were reared through to the adult stage and identified.

The findings were remarkable. On wild plants and in cropping systems, the wasps provided significant natural control of these pests throughout the season. Clearly, these parasitic insects were an important economic factor that we determined should be incorporated into management systems of agricultural crops.

Two of the wasps stood out in particular as they were the most abundant throughout the season by far on both wild plants and crops such as cotton. I learned that both wasps were in the family *Braconidae*, with the scientific names: *Microplitic croceipes* (Cresson) and *Toxoneuron nigriceps* (Viereck) *(=Car-*

---

5   Though these parasites are technically wasps in the Order, Hymenoptera, along with the other wasps and bees, they do not sting people, even when "handled." They are approximately the size of a house fly and the female does sting their caterpillar hosts for the purpose of injecting an egg, using a stinger-like structure, called an ovipositor.

Interactive life cycles of parasitic wasp and caterpillar host.

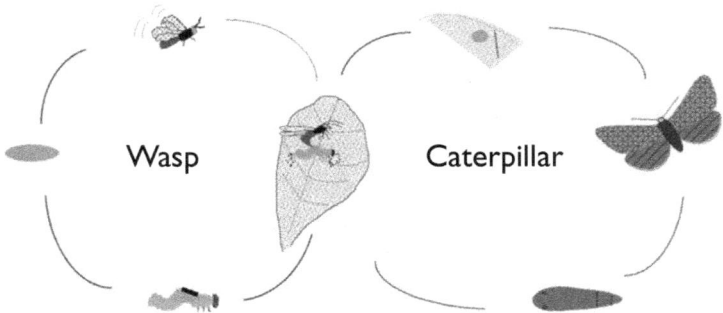

Cycling clockwise wasp deposits egg into caterpillar, which hatches as parasitic worm inside, eventually killing caterpillar. The worm chews out, forms cocoon. Wasp emerges from cacoon and cycle repeats.

Cycling counter clockwise moth deposits its egg on plant, egg hatches into caterpillar that feed on plant. Those that are not attacked by parasitoid mature to pupal stage. Adult moth emerges and cycle repeats.

diochiles nigriceps Viereck).[6] What was especially fascinating to me at the time was that these two wasps, though obviously of significant economic importance to agriculture, had received little research attention and so there was only a small amount of literature about either.

Of even greater intrigue was the fact that one of these wasps appeared to be very host-specific. *Microplitis croceipes* was found to attack and develop in both species of caterpillars. *Cardiochiles nigriceps*, on the other hand, was emerging at a high percentage from field-collected tobacco budworms, but never

---

6 Throughout my career this species of parasitic wasp was *Cardiochiles nigriceps Viereck*, which is the name used in all our publications. However, the taxonomic name has been officially changed to *Toxoneuron nigriceps* (Viereck).

from the cotton bollworm, even when collected from plants in the same field. What was going on? Was *Cardiochiles nigriceps* unable to locate the cotton bollworm? Were they locating both kinds of caterpillars but simply not stinging the cotton bollworm? Or was the cotton bollworm somehow immune to the egg and larval development of this wasp? The state of affairs provided an excellent model for understanding key mechanisms governing parasitic wasp/host caterpillar interactions, and would become the subject of my doctoral dissertation.

To me, it was an absolute delight to have the facilities and tools—some as simple as a binocular microscope—to pursue these questions, to provide me the ability to explore and study these creatures in ways that went far beyond the imagination of the young boy I once was in that cotton field. I felt like that same boy but with my curiosity now completely unbridled.

I could grow the different plants—cotton, tomato, tobacco, etc.—in our greenhouses or in field plots on the campus farm. Since we were growing both these caterpillars in our laboratory, I could place them on the plants in specific ways. I now recognized the *T. nigriceps* wasps, and knew how to find them as they hovered around caterpillar-infested plants in the field, particularly tobacco plants, and could follow them and watch their behavior when they were near the caterpillars. I could also catch some of the wasps with an insect net and bring them to the laboratory for study. I had dissection tools that enabled me to surgically open the caterpillars and I worked out a procedure to dissect a caterpillar that had been stung by this wasp and locate the tiny egg or larval stage inside. Other than two reports of field observations, one in the 1920s and the other from 1944, there was no literature about this wasp species. Until then, it had remained unstudied. As I probed through the dissected caterpillars and located the tiny eggs and larval stages for the first time, I realized, with a rather profound feeling, that I was likely the first human to see these stages of this wasp.

Through these field and laboratory observations, in cooperation with Vinson and Brazzel, I was able to confirm that this wasp was indeed host-specific to the tobacco budworm. I found that the wasps used chemical trails to track and sting the tobacco budworm caterpillars (similar to the way Daddy's fox hounds would chase the fox). These chemicals were much weaker in the cotton bollworm, so they seldom encountered these caterpillars. Further, through a series of studies using rather sophisticated immunological and chemical analyses methodologies, I found that when the cotton bollworms were stung, the egg and young larva were unable to develop because they were encapsulated by phagocytic cells. Inside the tobacco budworm, however, these egg and larvae were able to operate like a stealth bomber, never recognized as a foreign object. In other words, this wasp was highly specialized and very effective at attacking the tobacco budworm in all stages on all plants throughout the season. But in the process of this specialization, they gave up other hosts as a resource. It was like hunting with a rifle rather than a shotgun.

We published these findings in a combination of several papers in scientific journals that collectively stand as the initial groundwork of research on this wasp. Because of this work, the wasp, sometimes referred to as the "red-tailed wasp," is recognized today as an important pest control agent for the tobacco budworm throughout the southern United States.

Meanwhile, if the lab work was exciting to me, so was the work the entomology department was doing as a whole. We were growing. We were new enough as a separate department that we didn't always have the funds we needed, but we took care of things ourselves, improvising, and even building our own lab. Going back to my undergraduate work, I was rubbing shoulders with some of the brightest minds in the profession, some at Mississippi State, and others I met at various entomological conferences we attended in places like New Orleans, Atlanta, and Little Rock. People would come to these conferences from all over the world and it was always exciting to meet others in the field, all just as curious about new discoveries as I was. Being a scientist, I was beginning to see, was a bit like being a detective and we were all following clues and building on the new research that kept coming out. These conferences also gave me my first experiences in making oral presentations of our research work, helping to further lay the groundwork of my career as a professional scientist.

In New Orleans, at the national meeting of the Entomological Society of America, I met two professors from the University of California at Berkeley: Dr. Ken Hagan and Dr. Leo Caltagirone, two people I would later work closely with. Ken, in particular, would take me under his wing and mentor me. In fact, I would spend a large part of 1981 as a visiting professor lecturing at Berkeley. Overall, a whole new world was opening up for me.

During my time at MSU, my parents made a couple of visits to see me and were clearly proud. But the new world that was opening up to me was a foreign one to them. I sometimes envied my friends whose parents were more academically inclined, who understood their children's' pursuits, and who could engage with them on that level. My fellowship with my parents would always be within the context of their world, never within the context of my new world. But I loved them dearly, welcomed their visits, and was happy for them knowing that with Daddy's job with the highway department, they were now living comfortably in a house they owned.

While my professional life was opening up, so was my personal life. In June of 1965, just after I completed my master's, I married the girl I'd been dating since my sophomore year in college. Dianne Reeves had recently completed her bachelor's in music education. From Brookhaven, Mississippi, she was the niece of my high school typing teacher who had introduced us.

And things were only getting better. While working on my Ph.D., Dr. Brazzel gave me a research associate faculty appointment, a position that gave me a significant salary, benefits, even access to faculty housing where Dianne and I lived. I was making more money than I ever had before, more money certainly than my parents would have dreamed possible for me, and all while doing work that I loved. Because of the people around me, their ever-present encouragement and collaboration, and the opportunities to take advantage of early initiatives that were offered, I ended up finishing my Ph.D. at the age of twenty-four. Already, I had eight publications to my name. The future looked bright and, even before I'd finished my Ph.D., I found myself with a few intriguing job offers.

CHAPTER 7

A Great Time to be in Agricultural Research

"Hey, farmer, farmer, put away your DDT,
Give me spots on my apples, but leave me the birds
and the bees..."
—Joni Mitchell

"Mission completed!" I thought to myself as I exited the building on the west side of the Mississippi State University campus. It was the first week of September 1967, almost five years to the day since I'd arrived on campus and undertook this pursuit. The last two and a half hours, meeting with my doctoral committee to undergo the standard dissertation defense, had been grueling, but productive. It was the final step to completing the requirements for my Ph.D. The committee members had given me good marks and signed the approval page of the dissertation document. I was officially through. With a deep sigh—a mixture of pride and nostalgia—I stood for a while, gazing about the campus before me. "Oh, how I have enjoyed this place! And what a wonderful and productive time it has been. How could one pack more into a five-year period?" But all my objectives at MSU were now finished, and it was time to move on to the next chapter.

Conditioned on the expectation that I would complete these final requirements for the Ph.D. degree, I had accepted a position as Research Entomologist with the Southern Grain Insects Research Laboratory (later to be named the Insect Biology Population Management Laboratory; later still, Crop Protection and Management Unit), USDA Agricultural Research Service at the Coastal Plain Experiment Station of the University of Georgia (now named the Tifton Campus, University Of Georgia). I was scheduled to report to work the first Monday in October.

In addition to interviewing for the Tifton, Georgia position, I had interviewed in May and June for positions at the campuses of the University of Arkansas and Cornell University. I liked the people and environments of both. Cornell was especially tempting because of its stature as an Ivy League school. Plus, it's generally considered the birthplace of entomology as an academic field of study due to John Henry Comstock's work there starting in the early 1880s.

My trip to Cornell, in Ithaca, New York, was interesting to me for another reason: it involved my first-ever trip on an airplane. I enjoyed the flight into JFK, but then things got a bit confusing. I had a connection to make—Mohawk Airlines Flight 121 to Ithaca. But nobody was at the Mohawk counter and there was nothing posted about any Flight 121. I looked around and finally found someone who looked official and asked about my connection. Turns out the connection flew out of Newark. New York Airways ran a helicopter shuttle between airports in those days and I needed to take that shuttle to Newark and catch Flight 121 from there. All of this the man explained, but he did so in a strong Brooklyn accent that I heard through my Mississippi ears. He kept saying "Newark" and I kept hearing something that sounded a lot like "New York."

"You hafta go to Newark."

"Yes, New York. I'm in New York, right?"

"No, no...Newark."

"Yes, I know I'm in New York. I need to get to Ithaca."

"Newark!"

"New York."

On it went until the man finally pointed me toward a cab that would take me to the terminal where the commuter helicopter flew out of. "Just get on the helicopter," he said. Once on the helicopter I was apprised that the next stop was a place called "Newark" and it all finally fell into place for me.

Otherwise, the Cornell visit was wonderful, as had been the University of Arkansas visit. But I decided that the Tifton job was the one for me, the one that would allow me to do the research I really wanted to do. With a free month before I was expected to start, Dianne and I took a long drive up through the Mammoth Cave area of Kentucky, on to Niagara Falls and upstate New York, through Vermont and New Hampshire, and then back down through Boston and New York City, before driving through the Appalachians. Except for my interview at Cornell, I had never traveled outside the South and it was a wonderful opportunity to see these other parts of the country. Growing up as a baseball fan, I especially appreciated seeing Yankee Stadium and Fenway Park, two iconic ballparks that I made sure we visited. The sights all along the way were delightful and the month passed rapidly.

In Tifton, we found a small rental house. Our furnishings weren't much, so we decided to move ourselves. When we pulled out of our apartment at Mississippi State in the U-Haul truck I rented, I must have felt the need to make one last mark to punctuate my time at the school. Backing into the street at our planned departure time of 4:30 a.m., I turned one side of the front bumper into the neighbor's car, which I hadn't noticed was there the night before. The neighbors were a middle-aged couple who took great pride in their clean, smooth-running automobile and waking them at that time of the morning to report the news was not my favorite idea for a final farewell. Fortunately, they were very gracious about the matter and after exchanging the pertinent information and making plans for follow-up arrangements, we were finally off to Tifton.

The Tifton position was the perfect job for me for several reasons. From the personal side, it was within a reasonable distance of home. My parents could drive to see us within a day's time. Ithaca, on the other hand, would have required multiday travel by car or bus, or else travel by plane, none of which my parents could have handled by themselves. From the professional side, Tifton offered the subject area I wanted to work in—the study of beneficial insects in agriculture cropping systems and factors affecting their effectiveness as biological control agents. I felt that long-term safe and sustainable pest management strategies would have to be built from a research-based understanding of these natural enemy systems. That would be the central foundation.

It helped, too, that Tifton was located in the south, thus providing me with an ecological environment in which to work that I was familiar with. The arrangement of my employment provided another benefit. I'd be employed directly by the Agricultural Research Service arm of the U.S. Department of Agriculture, but located on the University of Georgia campus, working in cooperation with the University. Through an adjunct professor appointment arrangement, I could take advantage of all the benefits of the university, like having graduate students, obtaining grants, and gaining access to the school's facilities and equipment. It was the best of both worlds. At the same time I wouldn't have to spend an inordinate amount of time teaching classes. I could focus more on the research, which was my real academic interest.

As it turned out, for a young professional beginning his

career, I could not have selected a better backdrop of resources and cooperation. The Southern Grain Insects Laboratory, one of the largest facilities on the station, consisted of around a dozen scientists and at least twice that number of technical and support staff, including the full range of clerical, business, personnel, and farm and maintenance staff. The discipline areas of the laboratory included a mix of plant breeding, chemistry, insect ecology, insect pathology, insect physiology, agricultural engineering, biological control, insect rearing, and insect migration. The facilities were comprised of excellent laboratory, greenhouse, and abundant farm space, along with all the associated farming equipment. I found the morale, cooperation, and attitudes of the people within the laboratory to be superb. The overall Experiment Station consisted of well over fifty scientists plus technical and support staff covering a broad range of disciplines from entomology to agricultural engineering, chemistry, genetics and breeding (plant and animal), animal science (beef, dairy, and swine), soil and water, and plant sciences and cropping systems—corn, peanuts, cotton, forage and turf grasses, pecans, and blueberries, all with a full range of associated staff and equipment. There was a strong history and culture of cooperation between disciplines and state and federal programs. This availability of interdisciplinary resources and cooperators was important to me because I wanted to study beneficial insects on a total-cropping-systems level.

Some other more personal benefits to the job also soon emerged. Dr. Wendell Snow, who had been a fellow student and friend at Mississippi State, had joined the laboratory in Tifton a couple of years earlier, and had recommended me for recruitment to the director of the laboratory, Dr. Alton Sparks. Wendell continued to provide information and helpful personal and professional support for getting settled in the job and community. Our close friendship would continue throughout our careers. Dr. Sparks provided wise guidance and support. He also provided me with connections on the campus and within the USDA, Agricultural Research Service Agency, and the profession in general. He was also helpful in obtaining the necessary facilities and equipment for making my newly created position operational. Dr. Sparks helped me do nothing less than launch my career. I will always be indebted to him for recruiting me into the position and getting me grounded for an effective start. I could never have guessed back then that I would remain there my entire career.

A year after beginning my job at Tifton, another position became available at our laboratory for which I felt another fellow student from Mississippi State would be an excellent fit. Richard Jones had obtained his B.S. and M.S. degrees at MSU before transferring to the University of California, Riverside, where he was completing his Ph.D. I had an opportunity to pay forward Wendell Snow's kind support for me and I urged Richard to

Richard Jones (left) and I observing parasitic wasps in cage.

apply for the position and advocated on his behalf. Richard was subsequently selected and he and I shared many research interests and developed an excellent working relationship and an even closer friendship. Later, he moved to the Department of Entomology, University of Minnesota. Still, our research collaboration continued until full-time administrative roles dominated his professional time, as he went on to become the Department Head and Dean of Agriculture. Later still, he moved to the University of Florida, serving as Dean of Research and Director of the Experiment Stations for the Institute of Food and Agriculture. Nevertheless, our friendship continues to this day.

Even with all the assets offered to me, the truth is I was setting up something completely new, in a role where no one had worked before, ready to do research that had never been done. This was as overwhelming as it was exciting. There were no road maps to follow. There was freedom in that, but there was also great uncertainty. What methodology would I use, and what equipment and other resources would I need? How much technical help with what qualifications would I require? What were my timetables? What would success look like? Where would I even start? Those working around me had been engaged in ongoing research projects for years and had definitive goals, even daily or hourly goals. I had a blank slate.

Moreover, I was moving into my newly created position in a laboratory where all the other scientific programs had been established for some time. The pie had already been sliced, so to speak. All the office, lab, greenhouse, field plot space, and other such resources had been allocated. In the original planning of the laboratory, my position and program hadn't been anticipated. Space and resources would have to depend on other programs sharing and making room. And I knew that as my program grew, the needs would grow, too. The right mix of ambition, cooperation, and diplomacy was important. Looking back, I should not have been worried. The process went exceptionally well. I ended up the beneficiary of amazingly generous cooperation by all the other scientists, as well as wonderful coaching and leadership by Dr. Sparks, all of which I remain grateful for to this day. In addition to those not specifically named, these scientists included: John Young, Bill Mcmillian, Billy Wiseman, Neil Widstrom, Bob Burton, Phill Callahan, John Hamm, Edsel Harrell, Woodrow Hare, Deryck Perkins, Malcom Bowman, Tony Sekul, Bob lynch, Harry Gross, Dick Gueldner, Ron Myers, Dick Marti, Wayne Wolf, John West-brook, Sammy Pair, Larry Chandler, and Charlie Rogers, Glynn Tillman, Baozhu Guo, Tom Hendricks, Ben Fraelich, Carroll Johnson, Ted Webster, Richard Davis, and Patricia Timper.

In a general sense, this period of time provided abundant professional opportunities, but daunting challenges. Meaningful change was going to take time. On one hand, DDT and the flood of subsequent broad-spectrum, synthetic pesticides had been immensely successful in the short term.[7] Not only did they produce spectacular results, they were easily deployable and cheap and effective in small quantities against a wide array of pests. These pesticides resulted in the elimination of malaria-carrying mosquitoes and other health pests from entire countries. Consequently, from the mid-1940s until the early 1960s, they became *the* pest control tool in agriculture, health, and other arenas, seemingly to the exclusion of other methods. In fact, these chemicals brought inestimable benefits to humanity by providing immediate solutions for many widespread agricultural pest problems. Indeed, they did nothing less than relieve human suffering. Their success led toward automatic "cradle-to-grave" spray programs for agricultural cropping systems and the manufacture and deployment of these chemicals became a gigantic multibillion-dollar industry.

---

7   Initially, insecticides were the predominantly used pesticides in agricultural crops, but later herbicides would take a greater share of the usage.

The use of these pesticides also fit the larger dynamic of our times: the get-big-or-get-out, fence-row-to-fence-row mentality, partnered with the mindset that there was nothing we couldn't accomplish on a grand scale. Farms were grand, farm machinery was grand, and technology was grand. With our increasingly large tractors and associated equipment and chemicals, we now could plant, till, and harvest immense acreages of single crops. And while these large monoculture practices reduced diversity, disrupting nature's balance and fostering pest outbreaks and increasing erosion and loss of organic residue, etc., we could now kill off pests and correct soil issues with our newly developed, man-made chemical pesticides and fertilizer inputs. Postwar advancements seemed to show that man could bend nature to his will.

Except, of course, that he could not. The problems of widespread, indiscriminate uses of DDT and other pesticides began to manifest themselves in four significant ways. First, there was the inevitable resistance of the targeted pests to the pesticides, thus requiring a treadmill of higher and higher dosages. Second, there were induced outbreaks of secondary pests that had remained below damaging levels until the pesticides disrupted the balance by killing off their natural enemies. Third, there was the obvious contamination of the environment with toxic residues that reached up into the food chain. Nature had no means by which to break down these newly introduced synthetic compounds. They'd get into rivers and streams where they'd be picked up by microorganisms that bigger organisms would feed on, and then bigger and bigger organisms, on up to birds and rodents and other mammals. This was the build-up in the food chain that Rachel Carson had documented so well. And fourth, perhaps the most inevitable development of all, due to the build-up of resistance and disruption of pest-natural enemy balances, there were resurgences of pests at even higher numbers than before. This was the very thing the pesticides were developed to avoid, meaning that even if the pesticides could be shown to be ultimately harmless to humans, the fact is, over time, they simply weren't effective.

So, where did we stand with this state of affairs? *Silent Spring* had hit a nerve and public investments into programs to develop and implement safer, environmentally friendly alternatives were under way. My new job was one such example. But given the widespread gravitation to the pesticide model that had dominated for two decades, these programs would have to be built from the ground up. That would take time. For many, the problems were seen only as an issue of contamination of the environment, with the solution directed primarily toward a need for softer and more selective pesticides. But such a view failed to address the broader issues of resistance, balance of nature disruptions, and resurgence. Truly ecologically based redirections to pest management strategies using built-in governing mechanisms would be required for sustainable solutions, using the "other road" examples such as biological control with natural enemies, sex pheromones or other behavioral chemicals, sterile insect technology, or resistant plants. Some of our scientific leaders had recognized the complications years earlier and were already implementing redirections toward these programs.

I found inspiration in several cases as guiding examples. One highly innovative solution to insect pest problems had been recently employed to deal with the screwworm fly, *Cochliomyia hominivorax* (Coquerel), a species of fly that attacks the wounds of warm-blooded animals. This fly had recently proliferated throughout the South, from Texas on through Florida, and had become a menace of huge proportions. The female would deposit its eggs into the open wounds of animals, especially around pastures rich in livestock. The eggs would hatch and the larva would feed on the wounds, causing massive infection and death. Entomologist Edward F. Knipling conceived the idea of using sterile insects for population suppression and eradication. He observed that male flies of this species would mate repeatedly, while female flies mated only once in their lifetime. He reasoned that if sterilized male flies could be produced in large numbers and released into the wild population, they would essentially overwhelm and breed the screwworm population into extinction. After developing methods to mass produce the flies and sterilizing them with radiation, an experiment was carried out by Knipling and a team of other scientists on the island of Curacao off the coast of Venezuela where the screwworm population was soon eradicated. When it was replicated in the state of Florida and eventually throughout the United States, the screwworm population was effectively finished off by 1966.

Knipling's efforts did not go unnoticed. He was presented the National Medal of Science and was made Director of the Entomology Research Division of the USDA Agricultural Research Service. Knipling was in that position when I first began my work. Later, on January 11, 1970, the *New York Times Magazine* would proclaim, "Knipling…has been credited by some scientists as having come up with 'the single most original thought in the 20th century.'" The sterile insect technique is credited as being one of the most significant peaceful applications of nuclear radiation for the benefit of mankind. Knipling would go on to be decorated with numerous additional awards including the World Food Prize, Japan Prize, FAO Medal for Agricultural Science, President's Award for Distinguished Federal Service, and several honorary doctorates. He would gain a strong personal interest in my work with parasitic insects and we would develop a close working relationship, spending extensive time exchanging ideas on numerous occasions. I was honored to accept his request to write the foreword to a book he wrote on parasitic insects—*Principles of Insect Parasitism Analyzed From New Perspectives: Practical Implications for Regulating Insect Populations by Biological Means.*

As for biological control with natural enemies, one approach was termed importation. This involved the discovery, importation, and establishment of exotic natural enemies with the hope that they would suppress a particular organism's population. This had been most successful in situations where a pest had moved into, or been transported to, a new environment, usually without the natural enemies in place that would otherwise regulate its population and prevent major outbreaks. A couple of spectacular cases had helped foster a strong emphasis on biological control at the University of California. The introduction of *Icerya purchasi* Maskell, commonly referred to as the cottony cushion scale, had at one time nearly wrecked the California citrus industry. This pest, which gets its name from the white, cottony secretions is produces, was apparently introduced accidentally into California from Australia. In a hallmark case of biological control, *Rodolia cardinalis* (Mulsant)—the vedalia ladybird—a natural enemy of the cottony cushion scale, was discovered in Australia, where the pest itself came from. Several shipments of the vedalia ladybird beetle were released into the citrus groves of California and the results were stunning. Within a couple of years, the pest status of the scale insect was solved.

A second success story in California occurred with a weed pest. Klamath weed or St. John's wort, *Hypericum perforatum* L., is largely known as an ornamental and medicinal herb. Native to Europe, Africa, and the Middle East, this aggressive weed arrived in North America with early settlers. Unfortunately, Klamath weed also induces photosensitivity in livestock that feed upon it, leading to skin damage and related ills. Wild populations of Klamath weed spread over large acreages of California grazing land, sickening cattle and sheep in the 1940s. The beetles *Chrysolina hyperici* (Forster, 1771) and *C. quadrigemina* (Suffrian, 1851) were first released in 1945 and 1946, respectively. Although both species established, *C. quadrigemi-*

na proved especially effective for Klamath weed control. Within ten years of the first releases, the weed no longer threatened the livestock industry.

These classic lessons and other successful cases of biological control had not been lost on the University of California. Berkeley, Riverside, and Davis all created programs dedicated to biological control. Years later, as a visiting professor at Berkeley, I would work with researchers there. In addition to Ken Hagan and Leo Caltagirone, I would also work with Carl Huffaker, the primary leader of the Klamath weed program. But the biological control work that was being done at UC was the exception, not the rule. For the rest of the country, when pests became a problem, the default position was to create a "better" pesticide.

These case examples illustrated to researchers like me that investments in research and development could deliver highly successful alternative approaches to pesticides. I would be working with insect pests of row crop systems of Georgia and other Southern states, including cotton, corn, peanuts, soybeans, and vegetable crops, with a central emphasis on cotton. The pests of these crops, particularly those associated with cotton, were in a large way connected to the infamous boll weevil, the menace for which my Daddy had dusted with toxaphene, and for which an entire laboratory had been built at Mississippi State University.

The boll weevil, from Mexico and Central America, was first spotted in Texas in 1892 and had been wreaking havoc with the southern cotton belt ever since. The boll weevil was what we called a primary pest of cotton. It was the weevil's only host plant. Overwintering in a state of hibernation called diapause, under trash and leafy areas around the edges of fields, the weevils would come out each growing season, specifically attacking the fruiting part of cotton. They reproduced prolifically, usually requiring insecticidal treatment quite early in the season. Though the boll weevil had become resistant by the late 1950s to toxaphene and other chlorinated hydrocarbon insecticides like BHC, a new generation of insecticides called organophosphates became highly effective. Given this new group of insecticides, controlling the boll weevil was not the most difficult part. The major problem was that the treatment triggered secondary pest outbreaks by killing the natural enemies that otherwise kept these secondary pests in check. Further, some of the secondary pests were showing resistance to the insecticides. To make matters worse, there was the ongoing concern as to when the boll weevil would become resistant to the organophosphates, just as it had to the chlorinated hydrocarbons.

Resistance and secondary pests were two of the four major problems with pesticides I mentioned earlier. At the time I was beginning my work in Tifton, the "treadmill effect" of resistance had gotten so bad that seventeen to twenty insecticide treatments per season were being applied to cotton in Georgia. Around that time, it was estimated that forty-one percent of all insecticide use for agriculture in the United States was applied on cotton, directly or indirectly in association with the boll weevil. In addition to the severe environmental hazards, it was becoming economically impractical to produce cotton in the face of such massive insecticidal requirements. The cotton industry in Georgia and the entire South was being decimated. The cotton acreage in Georgia was 5.2 million acres in 1914, a year before the boll weevil's 1915 arrival to that state. By 1983, the cotton planted in Georgia was reduced to a low of 115,000 acres.

The secondary pest problem I always likened to an experience I encountered on more than one occasion as a youngster. Walking by a neighbor's house on a gravel road, I'd see their big, mean dog sound asleep on the front steps and feel certain that I could easily slip by without awakening him. But in the midst of my passing might come the neighbor's other dog, a little dog, running toward me yapping and yipping and make a loud ruckus. I was never concerned with the little dog; he was too small to do any damage. The danger was him waking up the big dog. The boll weevil was the little dog, controlled with the organophosphates. The problem was that the treatment woke up the big dogs.

Fortunately, major relief was on the horizon. The investments in research at the Boll Weevil Research Laboratory at Mississippi State and other locations were paying dividends. Through a detailed understanding of the boll weevil's year-round life history, biology, and behavior, a multi-tactic strategy for eradication of the pest from the United States cotton belt had been assembled and was being pilot tested. The package consisted of three major components: first, a highly effective trap for the boll weevil to better monitor their presence at low levels. Jim Tumlinson, a former fellow graduate student who would become an extensive collaborator, had identified the boll weevil pheromone. Subsequently, a commercial formulation of this pheromone, named "grandlure," had been developed along with an efficient trap design. Second, a combination of cultural practices consisting of early planting, use of early maturing varieties, together with post-harvest plant and stalk destruction, would be employed to minimize the number of boll weevils going into and surviving overwintering. Finally, every field of cotton with boll weevils would be treated with malathion in accordance to a prescribed schedule. Working with the state governments and through grower referendums, all cotton growers would be required to participate.

The eradication program would be implemented under the leadership of the Animal and Plant Health Inspection Service (APHIS) of USDA. Interestingly, my former major professor, Dr. Jim Brazzel, had been recruited by USDA-APHIS where he would provide the technical and operational leadership of the boll weevil eradication program.[8]

Eradication of the boll weevil would give us the opportunity to take the "other road" with the cotton cropping system. In the absence of insecticidal treatments for this primary pest, we also calculated that we would be able to manage secondary pest issues in an ecologically sound way, too. We could not afford to miss this opportunity to eliminate such outbreaks of pests like the corn earworm, tobacco budworm, fall armyworm, beet armyworm, loopers, aphids, and others. These were indigenous pests that had not been introduced into the area while leaving their natural enemies behind. I had seen from my graduate student research, and from earlier years of my life, that these pests had an abundance of natural enemies that often kept them under control. Yes, part of the problem had been the triggering disruptions from treating for the boll weevil, but there were some reliability factors that we did not yet fully understand. In some fields, outbreaks would occur. In others, even in close proximity, they did not. What were the differences in these situations?

For me, it was a matter of confidence in the power of nature. I witnessed numerous interventions with an insecticide treatment before the natural enemy had time to correct the balance. By waiting even just another day or so, the problem could have been solved naturally. But dispensing such advice required certainty. I needed an understanding of governing mechanisms based on solid biological knowledge. The same forces that allowed the vedalia ladybird to bring the cottony cushion scale into control could work for us; we had at times seen it in action. The players were in place. What was needed was a lot

---

8   The boll weevil eradication program did take place starting in Virginia and North Carolina in 1978, then moving westward and southward across the cotton belt. The eradication effort was highly successful, resulting in the boll weevil being eradicated from the United States except for areas in Texas near the Mexican border where an ongoing containment is required. The eradication is considered one of the world's most successful implementations of pest management.

more analysis and experimentation. It was a great time, in other words, to be a researcher.

CHAPTER 8

Fields of Wonder

It was April 1993. We stood peering into the wind tunnel like kids gazing into a candy store window. Doctors Jim Tumlinson and Louise Vet and I watched as the small red and black wasp crawled to the open end of a glass vial on the right end of the tunnel. About a meter away to the left stood a six-inch branch of a cotton plant, the snipped end mounted securely into a small cartridge of water. A small bollworm caterpillar had been feeding on the plant for the previous twenty-four hours.

Louise Vet (middle), Jim Tumlinson (right), and I observe activity in flight tunnel.

My technician, Thoris Green, was conducting an observation that she had repeated many times. She had just placed the parasitic wasp (a female) into the tunnel and closed the door. A steady breeze of air was flowing left to right in the tunnel, and so any odors being emitted by the target feeding site would now be passing directly over the wasp. As the wasp reached the top of the vial, she began to twitch her antennae, sampling the air. She took a couple of steps to the left then right, paused for a bit, twitched her antennae again, clearly oriented her body upwind, raised her two front legs in what we had come to call a take-flight pose, then lifted off. Once airborne, the wasp continued to remain oriented upwind, toward the cotton branch. She zigzagged left and right, apparently determining the path and bounds of the odor plume, then proceeded in a straight line toward the target again. About halfway there, she repeated the zigzag movement, more narrowly this time, seemingly re-establishing the bounds of the plume, then took a straight line forward. A few inches in front of the damaged leaf, she hovered in a holding pattern for a few seconds, then landed directly on the leaf and caterpillar area. Shortly after detecting the larva, she went into attack posture, coiling both antennae into a c-shape, raising her two front legs, and leaning backward. Then she lunged forward, thrusting her ovipositor into the caterpillar. We knew at that moment that an egg had been injected. The egg would develop into a wasp larva that would grow inside the caterpillar, kill its host, pupate, and become a wasp.

"Wow!" exclaimed Louise. "Remarkable. I never get tired of seeing this."

"Once the variables are all in place you can see the positive response coming as soon as she gets to the top of the vial," said Jim.

I agreed: "It's so great that we know the key parts. Remember when they would respond sometimes and other times just go to the top of the tunnel? We had no real handle on what was going on."

What we had just seen was the perfect example of a wasp's ability to detect and attack its caterpillar host. But we knew from our research that it represented only part of a larger whole. Imagine the complicated dynamics under natural field conditions. In order for a tiny wasp like that one to reproduce, it needs to find a certain type of caterpillar that feeds only on certain kinds of plants, and this may vary from one time to another and from one generation to the next. Consider the massive amount of foliage, space, cracks, and crevices caterpillars have available to elude the wasp, even in a single cotton field. The stakes of this cat-and-mouse game are high on both sides of the equation. Every generation of wasp must find adequate caterpillars in order to reproduce. At the same time, enough caterpillars have to escape attack for their own survival and reproduction.

For biological control, the ability of a natural enemy to find and attack enough pests to limit their numbers is critical. And so I wanted to know how this system operated. For that reason, I had selected this as a central topic of my program at Tifton when I'd arrived there in late 1967. I had also selected parasitic wasps of caterpillars the primary model for our studies.

That April day in Tifton was a day of celebration. During the previous ten years, we had worked closely with a team of excellent cooperators, especially our Dutch colleagues, specifically doctors Louise Vet and Joop van Lenteren, and their students of the Agricultural University, Wageningen. We all shared a strong interest in the subject and had recently made several key breakthroughs as to how parasitic wasps find their caterpillar hosts. In fact, a summary of these findings, by Jim, Louise, and me had just appeared in the March 1993 issue of *Scientific American*. Prior to that, we'd achieved the honor of publishing three papers in *Nature* and two in *Science*, a rare accomplishment. The standard of acceptance for a research paper to be published in either of these eminent journals is that the findings being reported not only represent a fundamental breakthrough in the specific subject discipline, but for science in general. We had reason to be proud. What we had learned about the interaction of plant, caterpillar, and wasp represented a level of sophistication far beyond our initial imagination.

How did we unravel the puzzle? How did we learn the secrets of the cat-and-mouse game playing out every day in cotton fields? Back in 1967, I knew there was a lot of ground-work to cover. The first several years were devoted to building the foundation of the program. I knew I'd need to know more about the year-round abundances and distributions of the various parasitic wasps of our cropping systems. I needed to establish laboratory colonies and rearing procedures for the chosen species of wasps in order for us to conduct replicated research. We needed to develop an understanding of their complete life cycles, learn how to handle and work with them under captivity, determine what type of cages to hold them in, how to provide them food, and what temperature, humidity, and light/dark cycles to employ. We needed a way to ensure mating and sex ratios. We had to know how to handle the parasitized caterpillars until the pupal stage of the parasitic wasp could be completed. After all that, we needed to figure out what methodology to use to study such matters like the wasp's host-finding behavior under laboratory and field conditions. Up until that point, essentially none of this had been done for the parasitic wasps we would be studying and very little for any parasitic insects. We were starting with a rather clean slate—a lot of opportunities, but a lot of work to be done before we could get close to the heart of how parasitic wasps find their hosts.

Richard Jones joined me early on and we worked together on some of the first breakthrough findings, identifying chemicals, called kairomones, used by parasitic wasps for close-range location and recognition of host insects. Our findings resulted in a publication in *Science* of the first ever kairomone chemically identified. Also working with us in the earliest years was Don Nordlund, a young scientist I hired who had recently completed his master's degree at the University of Georgia. Don proved to be a wise choice and was of immense value in getting our program effectively launched. The three of us, along with other collaborators, published prolifically and established ourselves as early pioneers in the basic biology, life histories, and ground-work for methods of laboratory and small-field plot handling of parasitic insects. We also published many of the earliest demonstrations of the role of chemical cues in the close-range, host-finding behavior of parasitic insects, and physiological/ immunological interactions during egg and larval development inside their hosts. We weren't alone. During this time, numerous similar programs were getting underway throughout the Unites States and internationally. But the secrets of how parasitic insects found and made use of the chemical cues we had discovered remained elusive.

Don, Richard, and I put together a book—*Semiochemicals: Their Role in Pest Control* (John Wiley and Sons)—to review the status of all the research being done at that time. Prior books on behavioral chemicals of insects had focused on sex pheromones and similar chemical cues associated with the behavior of the pest insects themselves, as opposed to the pests' enemies. Our book highlighted the shift in thinking, a shift born of the desire to find natural solutions to the problem of crop pests.

Around this time, Jim Tumlinson, whom Richard and I had known at Mississippi State, joined us. Jim, who was majoring in chemistry with a minor in entomology, had been a student in an entomology class I had taught there as a graduate student. Jim was now with USDA-ARS on the University of Florida campus in Gainesville and had become known for his work in identifying the boll weevil sex pheromone. Richard, in the late 1970s, relocated to the Department of Entomology, University of Minnesota, where he continued his cooperation with us, but was soon drawn into administrative duties as department head and then dean of the college of agriculture. Don, in the mid 80s, relocated to the USDA-ARS location in College Station, Texas with plans to complete his doctoral degree, but was soon offered an opportunity to move into a leadership role for the agency, managing intellectual property matters.

Back in Tifton and Gainesville, the collaborative work by Jim and me, focusing on how parasitic wasps find caterpillars, was gaining international recognition. As a result, we were drawing a stream of graduate students and post-doctoral scientists who wanted the experience of working with us for a year or two. The fact that our labs provided a combination of behavioral and chemistry perspectives had a strong appeal. Besides visitors from around the U.S., we had visitors from Chile, Brazil, Columbia, Germany, France, Poland, Israel, Sweden, Switzerland, Japan, the Netherlands, England, and elsewhere. In time, some forty scientists from fifteen countries would participate and make contributions to our program. Perhaps it was our joint Mississippi roots, but for whatever reason, Jim and I shared a synergy that, for years, fostered a gratifying spirit of cooperation and the free sharing of ideas between us.

By that day in April of 1993 when we watched the wasp find her host in the wind tunnel, Louise and Joop, along with their students, were playing a central role. Our earlier work had been constrained somewhat by the supposition that the wasps were hard-wired to track fixed chemical trails to locate their hosts. We assumed, in other words, that they operated primarily on instinct. Indeed, we had demonstrated such responses and identified the kairomones that triggered them. We observed the intense excitement exhibited by the wasps at the presence of these chemicals, including the intense use of their antenna (antennation). The kairomones were found to be present in the host feces, silk, and oral secretions (collectively called "frass"), as well as on the surface of the caterpillar itself. But these were nonvolatile chemicals, meaning they didn't vaporize; they didn't, in other words, create a scent that could be detected by the wasps in the air. They could only be detected at close range, such as when the material was placed on filter paper and the wasp was allowed to walk over it or antennate it.

So how in the world do wasps find their hosts in the midst of a busy cotton field? Surely, they can't just fly around randomly looking for caterpillars hidden among the cotton plants. We knew there had to be something else in play. I had observed wasps hovering around plants in cotton fields, clearly detecting airborne odors. We needed to determine how the longer airborne searching worked, and it struck us that we needed a bioassay system—an analytical method to study the flight responses of wasps to volatile chemicals under controlled conditions. That's where the wind tunnel came in. Mike Keller, a post-doctoral student working with me, provided one from a project he'd worked on earlier. The wind tunnel was designed to maintain something called laminar flow—basically a smooth layer of evenly distributed air that could be passed from one end to the other at desired speeds, and environmentally insulated from extraneous odors and air currents.

The wind-tunnel system turned out to be effective. Our standard procedure was to introduce the target odor source material on the upwind end of the tunnel, and release the wasp at the other end. The standard target consisted of a portion of a cotton, cowpea, or corn plant with the feeding caterpillar, wounded area of plant and residue frass still present. This combination of materials is referred to as the PHC—plant-host complex. Not long into our experiments, we constructed a second wind tunnel about twice as big as the first and suitable for observing patch-size searching behavior through the placement in the tunnel of multiple potted plants.

Through a series of outstanding individual and team studies

involving all of our players, we began to successfully unravel nature's secret. We chose four different parasitic wasps of the family *Braconidae* for use in the studies: *M. croceipes*, *Cotesia marginiventris (Cresson)*, *T. nigriceps*, and *Microplitis demolitor* Wilkinson. As we proceeded, we discovered that the processes operated the same with all four species. The plants we used were cotton, cowpeas, and corn. Although the blend of chemical emitted by the plants varied, we found the response mechanisms to be the same for all the plants.

Caterpillar hosts feed on many different plants and different parts of the plants. In each case, the odors are different. Even the age of the plant will make a difference. This meant that wasps must be able to operate under vast field conditions across wide ranges and mixes of chemical cues, and yet somehow be able to determine which set of odors are associated with their hosts at any one time and home in on that particular mix. Yvonne Drost and Oliver Zanen, working in our Tifton lab, and Fred Eller in Gainesville, showed that after antennation of feces from caterpillars feeding on cowpeas or cotton, the wasps then flew better to the respective feces antennated. There was, that is to say, a degree of *learning* that was taking place. But how?

Answers to questions like these come about in interesting ways—sometimes gradually, at other times in a sudden burst. Jim and I were on a trip to Holland around this time. One evening we were sitting out on the front porch of the Agriculture Center at the Agricultural University, Wageningen discussing this subject when a joint idea emerged that we felt certain could give us the answer. We quickly planned a study to test our idea, which we conducted immediately upon our return home. The results of that study, entitled, "Host Detection by Chemically Mediated Associative Learning in a Parasitic Wasp," were published a few months later in the January 1988 issue of *Nature*.

What we were able to show was that when the wasps encounter and antennate the host caterpillar feces, they learn to recognize and subsequently fly to the mix of volatile odors present at that time, even novel and previously unattractive odors. They accomplish this by linking these volatile chemicals to a nonvolatile host-specific recognition cue. By so doing, the wasp is able to confirm that the feces in question are, indeed, from the host. For the wasp, it's an associative learning process. We were able to extract the nonvolatile chemicals with water and, by placing a bit of this water extract on filter paper, we could actually train the wasps to fly to any novel volatile odor we desired. At one point, we used vanilla. We let the wasp antennate the filter paper while smelling vanilla and, to our delight, the wasp would then fly to the odor of vanilla.

Further, we found that this linking occurs in a similar way when the wasp stings a caterpillar host to deposit an egg. Essentially, the wasp uses a "tasting while smelling" process. During these encounters of either antennating the feces or stinging the caterpillar, the wasp detects the host-specific chemical(s) with taste receptors located on the antennae or ovipositor, thus confirming for the wasp, "Yes, this is a host." This recognition is innate, not learned. But at the same time, olfactory receptors located on the antennae sample the air to determine what odors are associated with the host site. *That* recognition is learned. Afterward, the wasp will then fly to those odors. Through the use of this system, the wasps are able to maximize their success by constantly updating their responses to coincide with what the host caterpillars are feeding on and how these feeding sites smell. They can then home in on those sites.

The ability to learn associated odors by linking their presence to a recognition cue in the feces without actual contact with the host larvae is a beneficial adaptation since the caterpillars are often in crevices or otherwise unreachable during a particular visit by the wasp. Recognizing the host and linking the volatile cues allows the wasp to maintain accurate tracking cues, perhaps prompting her to pursue further examination, maybe probe nearby crevices with her ovipositor more persistently. Hans Alborn, working in Gainesville, demonstrated that this host-recognition chemical was, indeed, of caterpillar origin and not diet related.

Felix Wackers, working in Tifton, then found that the wasps supplement the use of odors by the learning of *visual* cues, including color, patterns, and shapes associated with host encounters. For example, if we placed color cues such as black and white versus solid orange squares at sites where hosts were encountered, the wasps would soon learn to prefer the color cue associated with the host caterpillars. Although responses to odor alone were always stronger than visual cues alone, the combination of odor plus visual cues was additive. This ability greatly enhances the foraging success of the wasps. If, for instance, a majority of caterpillars located in a cotton field is feeding on white blooms on one occasion and on triangle-shaped squares on another, the wasps will quickly adapt for each occasion, making note of the pertinent visual cue.

Given the importance of this associative learning process in host-finding, the question naturally arose as to how the wasps find their very first hosts. We found that odors associated with a particular plant being fed on by a host caterpillar become woven and impregnated into the cocoon as the mature wasp larva forms that cocoon. Franck Herard, working in Tifton, showed that wasps that emerge naturally from these cocoons fly well to such plants infested with caterpillars. Conversely, when he cut the wasps in their pupal stage from the cocoons to complete their development, they responded poorly to caterpillar-infested plants. The responses significantly improved when he then allowed them to walk on and antennate the cocoons. Clearly, the young wasps rely on the information gleaned from their cocoon as their best bet in searching for their first hosts. Thereafter, they update their information based on their subsequent experiences.

Keiji Takasu, a visiting scholar from Japan, wondered, upon his arrival to our Tifton lab in 1990, about food sources for the adult beneficial wasps. We knew that the adult wasps were nectar feeders, but the questions remained—how did they find this food and how was that tied in with the extensive travel activities required for finding host caterpillars? By evaluating the wasps' flight tunnel responses after feeding them sugar water while simultaneously exposing them to certain odors, some remarkable answers began to emerge. Keiji found the same kind of learning as with odors associated with encounters with caterpillar hosts. The wasps quickly learned and subsequently would fly to novel odors such as vanilla and chocolate. But these studies revealed much more. The taste receptors for food quite naturally are located on their mouthparts, as opposed to the antennae or ovipositor as is the case for hosts. Via these different taste receptors, odors detected by the same central olfactory system are separately linked to the wasps' respective appetitive needs (e.g., host versus food). Each odor then serves as a cue for locating the respective resource to which it is linked. Furthermore, Keiji found that wasps, so trained, subsequently responded to these learned cues in accordance to the relative level of need at any given time. For example, when wasps trained to link vanilla to hosts and chocolate to food were provided choices of these two odors in flight tunnel tests, hungry wasps chose chocolate and well-fed wasps chose vanilla. When the taste-to-smell pairings were reversed, a similar reversal in choice followed. We published these remarkable findings in *Nature* in 1990.

Of course this kind of association in mammals is well known. And we knew through our earlier experimentation that wasps had a surprisingly sophisticated olfactory system for detecting and tracking odors. But scents are inherently arbitrary. What makes them truly useable is the animal's ability to index the scents, to associate them with food or procreation or shelter or any other resource necessary for survival. We had never imagined a wasp, with a brain the size of a pinhead, having a foraging system—sophisticated enough to learn, integrate, store, and subsequently process information—that could rival that of higher animals.

Keiji went on to demonstrate how the combination of antennation of frass and oviposition in host caterpillars determined not only the learning of host cues but the length of their memory. As described above, the wasps learn and subsequently respond to odors associated with host frass alone, without an accompanying encounter and/or sting of a host caterpillar. However, he showed that an accompanying host sting (within five minutes) significantly extended memory and subsequent response to the learned information.[9] This system provides a way for the wasps to utilize the frass for detecting the presence of host caterpillars hiding in cryptic sites, while still requiring periodic encounters in order to continue the search at a given site. Through a combination of elegant procedures, Keiji showed that the key factor in strengthening memory was the contact of the hemolymph with receptors on the ovipositor.

We found more efficiencies. All the wasps of our study were solitary species, meaning that only one individual develops per host caterpillar. Therefore, going back over the same territory is wasted time for a wasp. Felix Wackers and Bill Sheehan, working together at Tifton, found that the wasps minimize such duplication in two primary ways. First, by leaving their odors on the surface where they've walked and on caterpillars that they've stung, thereby signaling that the space in question has already been "worked." Secondly, they maintain somewhat of a spatial map and, thereby, limit duplicate searching.

Our discoveries regarding the interactions between cater-

---

9   This work was initially begun by William Martin, a Post Doc working at our Tifton Lab prior to accepting a permanent position with a private corporation.

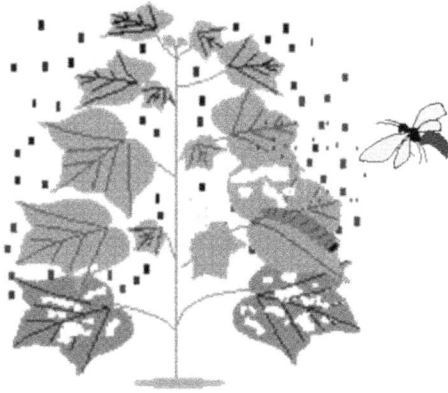

Plant Emitting Odor
Wasp Responding

pillar and wasp were nothing less than stunning. Concluding
that these wasps were not acting out of pure instinct, that they
were not hard-wired but were, instead, learning on the job, was
a surprise to all of us. But we weren't done being surprised.
As it happens, there is even more that goes on in your typical
cotton field. It turns out that the plant is far from a disinterest-
ed bystander or helpless victim of the caterpillar. Ted Turlings,
working in our Gainesville lab, revealed something   every bit
as amazing as the learning process of the wasp. Plants, being
fed on by caterpillars, draw help from wasps by producing and
emitting a set of chemical signals. These chemicals, called ter-
penes and sesquiterpenes, are more attractive to the wasps than
even the scent of the feces. It's the victimized plant's way of
sending up a flare or tapping out an SOS. The signals provide
a highly effective way for the wasps to quickly find infested
plants in places such as a cotton field with thousands of plants.
    Moreover, what Ted showed was that incidental damage,

replicated by cutting the plant with a razor, elicited only brief bursts of green leafy volatiles. But placing spit of caterpillars on the wounded area, thereby simulating attack, induced the plant to go into full SOS mode. Hans Alborn, of Jim's team, then isolated and identified the primary chemical in the spit triggering the plant responses, and named it "volicitin." Others of his team, including Paul Pare, Ursula Rose, and John Loughrin, went on to work out key details of this plant signaling system: 1) that these volatile compounds were synthesized *de novo*, 2) they were emitted from the whole plant with little lag time, and 3) that the volatile signals were emitted during the light periods, when the parasitic wasps are active, even though the caterpillars feed at night and day.

Consuelo De Moraes, working in the Tifton lab and later in Gainesville, uncovered yet another component to the plant-wasp dynamic. There are often many different kinds of caterpillars and many other insects feeding on plants throughout any given field. Each kind of caterpillar is a suitable host only for certain types of wasps, but not other wasps in the field. So, if the plant emitted the same emergency distress signal in response to all kinds of caterpillars, a wasp might find itself drawn time and again to a situation of little value to her. A false alarm. She'd eventually stop coming. So how does the plant let a wasp know that not only is it being attacked, it's being attacked by a caterpillar that would make the perfect host for the wasp? The deeper the well, the sweeter the water. The marvels of nature once again provided us with an answer. Using two very closely related caterpillar species and cotton and tobacco plants, Consuelo, working with others from our team, showed that the plants emit a different blend of chemicals depending on the species of caterpillar feeding on them. Furthermore, the wasp can distinguish between these different signals and respond specifically to the plants attacked by their host species. Hans Alborn, and others of Jim's team, showed that variations in the chemical elicitor, identified earlier among the different caterpillar species account for the distinctly different signal response for each plant-feeding species—in similar fashion to the bar codes for different items in the grocery market.[10]

The discovery of this remarkable ability of plants to recruit wasps to their defense when under attack by caterpillars provided answers for many of our previously unexplained pest outbreaks. A series of studies by Anne Marie Cortesero, Oscar Stapel, and Dawn Olson found that factors like soil nitrogen levels, water stress, and landscape management could substantially impact the plant signaling, and associated attributes in recruiting and supporting their natural enemy allies. So, these valuable plant attributes were being unknowingly weakened through plant breeding programs and certain agronomic practices such as soil management regimes.

Parasitic wasps of the caterpillar stage of pests were the primary focus of our studies. Yet, a wide diversity of beneficial wasps attacks other stages of the pest, including the egg stage. These eggs are attached to the plant by the female moths and hatch into caterpillars in about three days. Some of our work included studies of a group of tiny wasps of the genus *Trichogramma*, which are about the size of the period at the end of this sentence. These minute wasps commonly occur in the wild and help control the pests. They deposit their eggs inside the pest eggs, and go through their egg, larval, pupal stages inside the host eggs before chewing out as adult wasps. In keeping with our other studies, we pursued the question of how these tiny wasps traveled about and located these host eggs, which are only about the size of the tip of a pencil lead. Furthermore, these egg hosts do not leave chemical trails through feeding and

---

10 Around the same time, two of our Dutch colleagues, Marcel Dicke and Maurice Sabelis, conducted elegant parallel studies demonstrating similar phenomena with predatory mites and plant-feeding spider mites, indicating the likelihood of a widespread presence of this system in plant-natural enemy relations.

defecation, as in the case of caterpillars, and the tiny wasps are too small to navigate airborne chemical plumes upwind to the host site. However, Lucas Noldus, working in our Tifton lab as a visiting doctoral student from the Agricultural University, Wageningen, the Netherlands, designed and conducted a series of elegant studies revealing remarkable alternate ways that the wasps track airborne trail odors to locate their hosts. He demonstrated that the wasps locate the mating and oviposition sites of the adult moths by eavesdropping and tracking their mating pheromone and other trail cues. Additionally, rather than track odor plumes upwind, they drift with the wind until they detect the airborne odors, then land at those sites where they then follow close range trails by short flights within the canopy and/or by walking. Based on this and related research, Lucas completed his PhD from Wageningen in 1989 with a dissertation entitled "Chemical Espionage by Parasitic Wasps."[11]

The academic and research related impact of these breakthroughs published in the journals of *Nature* and *Science* were far-ranging. For example, Ted's discovery of plant signaling and Hans' identification of the eliciting chemical, were both published in *Science* and have been cited over 1,000 times each. Meanwhile, the doors to our labs were ever busy accommodating an insatiable appetite of science-related popular press around the discovery of how smart wasps and talking plants team up against sneaky caterpillar pests. Indeed, the times were exciting and the accomplishments fulfilling.

From a personal standpoint, our discoveries were extraor-

---

11   On an interesting personal note, while conducting his research as part of our team, Lucas adapted a handheld computer to serve as an event recorder by designating specific keys along with the timing, frequency, and sequence of those key strokes to correspond to given behaviors of the wasps. Recognizing a broad need and value of this methodology, he went on to become founder and CEO of Noldus Information Technology, which now provides solutions for behavioral research on a worldwide basis with twenty-three offices located in ten countries.

dinary to me in and of themselves. Nature, which I had held in awe since the cotton fields of my boyhood, was surprising me in ways I could never have imagined. The more I learned, the more amazed I became. Nature was revealing her secrets to us and I was beginning to understand the wondrous complexity of this magnificent drama, playing out daily in ecosystems everywhere.

From a more practical standpoint, but every bit as wondrous, is that the combination of these astonishing discoveries using the plant, caterpillar, wasp model was coming together to reveal a very key central point. These cotton crops and other such agroecosystems were not only sites for amazing feats of nature—powerful feats that acted to keep nature in balance, living systems made up of a network of tightly interwoven, interactive components—they were in essence macro-organisms. This appreciation of their holistic makeup was being lost through our trends of reductionist thinking and the interventionist mindset. Sustainable management of agricultural systems would require that we gain a better appreciation of this holistic perspective of agroecosystems. For example, managing a pest should always begin with the question of "Why is this pest a pest?" not "How do I kill this pest?" The answer for most such undesired variables is an adjustment of some aspect within the system, versus an external intervention such as a pesticide treatment. It was the significance of these discoveries that led to the Wolf Prize.

Our discoveries didn't end there. In 1998, Jim Tumlinson and I received a rather interesting guest, a representative from DARPA, the Defense Advanced Research Projects Agency. As it turns out, DARPA had a program called "Controlled Biological Systems." The guiding premise of the program was that natural systems could be a rich source of discovery and development of new innovations for military and defense purposes. DARPA took notice of our publications in *Nature*, *Science*, and *Scientific American* demonstrating how parasitic wasps exploit learned odors to locate caterpillar hosts, and how the wasps could even be trained to respond to novel odors in flight tunnels. DARPA wondered whether methodologies could be developed where wasps could be used like canines to detect explosives, drugs, diseases, and other items of military, medical, or forensic interest.

Their idea was certainly intriguing, but it seemed like a long shot. Nevertheless, after a series of discussions, DARPA agreed to fund a four-year project to be pursued by Jim and me at our respective labs. We assembled our team starting with a search for someone to head up the engineering aspects of the project. We didn't need to look far. The perfect choice for the role was Glen Rains of the University of Georgia who had already collaborated with us to help pursue his interests in the area of sensors for agricultural management. Glen shared our interest in sustainable agriculture, too. Others who would come on board, in addition to those we worked with before, included Dawn Olson, Moukaram Tertuliano, Torsten Meiners, Veronique Kerguelen, Claire Bonifay, Jeff Tomberlin, Marco D'Alessandro, Jelijko Jurjevic, and D. Wilson.

Our first step was to determine whether the wasps could even smell chemicals associated with the likes of explosives, nerve gases, diseases, or drugs. And if so, could they detect them at levels suitable for the purpose? Through a series of initial studies we determined that wasps possessed a surprisingly broad receptor system that was, indeed, capable of detecting all the chemicals of potential interest to us, and at very low amounts. Using the methods we'd already become proficient with, we quickly trained the wasps to associate any of the chemical odors with food or hosts. In the flight tunnel they would track the odor plume upwind and land on the source of the odor, typically a portion of filter paper on which the chemical was placed. The wasps trained to associate the odor with food would rotate around and examine the surface of the paper in a clear attempt to locate the food. Wasps trained to associate the odor with hosts would display the characteristic attack posture.

Based on these findings, we developed a system for using the wasps as a tool to detect targets of concern. We chose to use the association with food as the primary training method and developed a protocol for handling and training procedures. Upon emerging from their cocoons, the wasps were held without food for forty-eight hours. They were active and appeared healthy, but were obviously hungry and fed aggressively when food was presented. They were allowed to feed on a droplet of fifty-percent sucrose water for ten seconds while smelling whiffs of the target odor. This ten-second feeding was repeated three times with sixty-second rests in between. Wasps so trained were found to display positive responses to the target odor.

But just how to proceed with deployment of the wasps at actual test sites? The wasps weren't exactly domesticated animals that you could take around on a leash. Nor would they behave in a manner where you could release them in an area like at an airport. And what behavior would they display to report that a scent or other indicator of a target of interest had been encountered, and how would that behavior be measured?

We decided to containerize the wasps in a device that could be moved around directly by hand or remotely, while air samples were presented to the wasps. A device dubbed the "Wasp Hound" was developed for this purpose. The successful design of this tool required extensive work to understand how the complex interactions of taste, olfaction, and visual cues, along with the state of hunger and different handling protocols affected the airborne and crawling responses of the wasps to various light and air movement After a number of prototypes, and after extensive work to understand the interactions of taste, olfaction, and visual cues with state of hunger; airborne and crawling responses to various light and air movements; and the requirements of handling and training protocols, we settled on a device dubbed the "Wasp Hound." The design and operation-

Glen Rains (left) and I examine a device to deploy trained wasps as chemical detectors.

al process of the device are depicted in the figure below. This specific product was developed primarily through the creative work of Glen Rains and Samuel Utley, a graduate student working under Glen's supervision, but the string of prototypes and background work to get to that point emerged from the work of those named above and others.

As you can see, the cylindrical-shaped device (approximately six inches long and three-and-a-half inches in diameter) consists of a small fan, LED light, camera, and cartridge for holding the wasps. The device is designed to be connected to a computer and electrical source for powering and monitoring. The fan pulls a steady stream of air (four milliliters per minute) from the sample area into the wasp cartridge via a small hole at the center of the cartridge, then on through and out of the cylinder.

The behavior of trained wasps placed in the cartridge is monitored by the camera and transmitted to a computer for ongoing analysis. We found that five trained wasps was an optimal number so as to account for individual variability among wasps. Whenever the odor for which the wasps were trained was encountered, the wasps would promptly and briskly begin moving toward and concentrating around the air entrance port in an excited manner, obviously in expectation of locating food. A software program designed to provide an ongoing measurement and analysis of dark pixels concentrated in the vicinity of the air entrance port would signal a positive response within ten to twenty seconds.

An experimental model of the device interfaced with a small computer and used in a portable handheld manner was successful in several demonstrations including agricultural-related

A. Wasp Hound Device

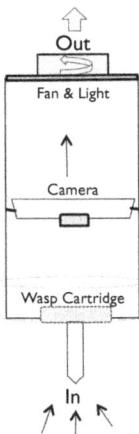

B. Camera View of Response

Negative Response
Random Movements

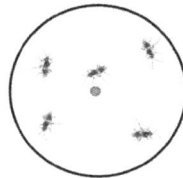

Positive Response
Concentrate Around Air Inlet

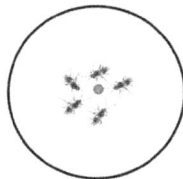

"Wasp Hound" detector system and how it works.

A. Five trained wasps are placed in the wasp cartridge where their behavior is monitored by a camera and connected computer system as sampled air is passed through.

B. Wasps respond as indicated when the targeted odors are detected, and a customized software program measuring concentration of dark pixels around the air inlet signals a positive response.

purposes such as the presence of plant diseases, or molds and similar concerns in stored products; forensic investigations for detection of illegal substances, buried bodies, and accelerants for arson; military of defense purposes such as explosives or other security concerns; and medical concerns such as monitoring for diseases. The sensitivity and reliability was demonstrated to be comparable to that of canines, and superior to any of the current electronic noses.

For us, the findings confirmed what we knew: no matter what it is, nature invented it first. It was just another example of the wealth of remarkable tools that can be harvested through understanding her. This exposition of the training and demonstrated use of invertebrates in a manner similar to the historical use of dogs brings an entirely new and vast potential resource into the picture. The downside is that, due to its completely novel nature, there is no existing industry to transfer this technology into use. One has to be developed. An incubation program such as a public-private partnership was explored for that purpose, but has not been achieved. Donnie Smith, Executive Director, Center of Innovations for Agribusiness and the Georgia Department of Economic Development were very interested, but an untimely back injury during that period precluded me from travel and meeting activity required for adequately exploring options they were pursuing.

All of the marvelous discoveries of how wasps interact with their environment had me looking at cotton fields in a way I never did as a little boy growing up in Mississippi. Even with the sense of wonder and fertile imagination I had as a child, I could never have conceived of the hidden activity that goes on within the rows of cotton plants—the caterpillars feeding on the leaves, the plants sending out distress signals, the parasitic wasps hovering around looking for clues. The interplay of all of the elements that make up this holistic, self-contained system was a stunning and beautiful discovery, and one, as I would learn, with larger ramifications. Today, I can't drive by a field of cotton or tobacco or soybean or corn without thinking about the extraordinary dynamic actively playing out within that field.

In the meantime, the discoveries were gratifying to me in another way. The collaborative process itself, with all of these wonderful scientists and researchers, was immensely rewarding. We were a group who gave freely to one another, sharing information and results without hesitancy, tossing ideas and theories and hypotheses back and forth. All told, we worked together for years, meeting for frequent discussions, going to quarterly retreats and overnights, enjoying dinners and fellowship together. And all the while remaining excited by the knowledge we were collectively uncovering. I cannot imagine a more selfless group of people to be associated with.[12]

And of course it wasn't all work, although the work would have been rewarding enough. It might come as a surprise to some that getting a bunch of scientists together at a retreat can be a recipe for some enjoyable hijinks. Like the one night at Tall Timbers Research Station, north of Tallahassee. A few of us managed to convince a fellow researcher, who will remain nameless here, that a fun nighttime activity, after a day of lengthy discussions about plants and insects, is to go into the woods for a little snipe hunting. As any Boy Scout who has spent any time at a summer camp can attest, snipe are completely fictional forest birds that, as we explained to our trusting friend, lose the ability of flight once the sun goes down and can be rounded up by driving them to lower ground, like a ditch, where they naturally run if chased. All our colleague

12 Jim and I had several other scientists participate in our program, including: M. Altieri, G. Morrison, T. Mueller, D. Whitman, W. Martin, J. Dmoch, L. Noldus, G. Prevost, W. van Giessen, Naoki Mori, P. McCall, J. Strong-Gunderson, A. Demata. Though their work is not specifically covered herein, they all made valuable contributions to the overall success of our program.

needed to do was stand in the ditch with a bag while the rest of us went through the woods to find the snipe and drive them toward him. Snipe, we asserted, make an excellent breakfast meat. Gamely our fellow researcher stood in a ditch holding the bag while the rest of us went back to the cabin to sit on the porch and sip a few beers. Eventually, he heard us laughing through the woods and ambled back to the cabin. "You almost got me," he said. We pulled the same trick on another researcher at a retreat at Fernandina Beach in Jacksonville with similar results and later presented the both of them with official certificates designating each as "a Member of the Royal Snipe Society."

Of course it was all in good fun, made possible by the kind of camaraderie born of our close working relationships and mutual wonder at the discoveries we were making. But through it all, we never lost sight of the original goal—to strengthen biological control and the "other road' as alternatives to pesticides. How could we take these remarkable findings and use them? Yes, they were fascinating in and of themselves—astonishing even. We had knowledge nobody had ever had before, and a foundation for sustainable agricultural initiatives. But now we needed to find a way to introduce it into mainstream agriculture.

CHAPTER 9

"Built-In" Pest Management

"You can't have any good guys without a few bad guys. That's fact."

So says Alton Walker. Alton and I have been friends since our days at Mississippi State where we went through our master's degree program at the same time. Also a native of Mississippi, Alton continued his education at Clemson University, obtaining a Ph.D. in entomology prior to his career in agricultural consulting and farming in Georgia. He and I came to have a shared interest in ecologically sound farming, and in the mid '90s we collaborated with a team of scientists on sustainable cotton production following the boll weevil eradication. Alton is a scientist with some skin in the game. He's pursued the application of his conservation/ecologically based ideas with cotton production on a 600-acre portion of his own farm.

As Alton will tell you, the common practice of cleaning a field down to bare soil after harvest and leaving it barren over the winter is a harmful practice for multiple reasons, including pest management as well as natural resource conservation. "Farming's been the victim of the advances of highly mechanized 'big farming' approaches," he says. "Through the use of large equipment like harrows, plows, and mowers, enormous

Alton Walker standing in a field with mixed cover crop where he practices whole ecosystem management

portions of biomass are removed from countless stretches of land. The land is then tilled and planted into monocultures from ditch bank to ditch bank. Then, mechanical cultivation and chemical pesticides are used to restrict diversity, while fertilizers and irrigation foster a lush growth of crops. Every year, the process starts over, meaning there's never an opportunity for a true, natural ecosystem to develop and remain in place for the length of time it takes for it to become balanced and efficient. It's no wonder pest outbreaks occur. On the other hand, perennializing the field—growing something year-round—helps promote a much more stable and balanced environment. We have to find our way back to approaching farming, including pest management, with an understanding of how to manage the ecosystem in which we live."

The team Alton and I collaborated with in the '90s was an interdisciplinary group of researchers that included Sharad Phatak, Rick Reed, John Ruberson, and Jim Hook, and Glenn Harris with the University of Georgia, and Philip Haney with my laboratory in Tifton. Eradication of the boll weevil, which had been completed in Georgia in 1990, and, later, essentially all of the United States, presented the cotton industry with a unique opportunity to advance sustainable agriculture. The eradication had been one of the greatest technical successes in agricultural history, with immense potentials in economic and environmental benefits. To completely eradicate the presence of a pest of this magnitude from the entire cotton belt! In Georgia, insecticide use was already dropping sharply, with average crop revenues increasing markedly. By 1995, the use of fifteen to twenty treatments per year had been reduced to three to five treatments. Grower interest in biological control and sustainable agriculture had never been higher, but a shift in thinking on when and how to give nature more time was going to be needed. The boll weevil had been an invasive pest without any effective natural enemies. Quick to reach damaging levels in early season, it was an especially devastating primary pest because the necessary insecticidal treatment for its control regularly spurred a sequence of secondary pest outbreaks. But now, for the first time, we could put in place an ecologically based management system without the disruptive influence of the early season boll weevil treatments.

In this new era, we could promote the adoption of cotton production as part of a healthy year-round landscape system, with approaches to pest management that deal with the natural enemy/pest complex being a vital part of that overall system.

But to take advantage of this new era, we knew there needed to be a lot of educational outreach to the grower community, including on-farm demonstrations with associated data. Otherwise, we could miss the opportunity and drift back to pesticides as the dominant pest-management practice.

Figure 9a below shows the conventional high intervention methodology (Box 1) as contrasted to year-round landscape ecosystem management (Box 2). The conventional, high-intervention approach has predominated cotton production and pest management for years, particularly since the advent in the 1950s of big farming. After harvest, the field is mowed and harrowed, rendered barren until spring when the process starts over. Because of this winter and early spring "wipeout" of everything prior to planting, the ecosystem—as represented by the typical "ecological growth curve"—is never able to achieve equilibrium status. So, there are no relays of natural enemy/pest

Figure 9a

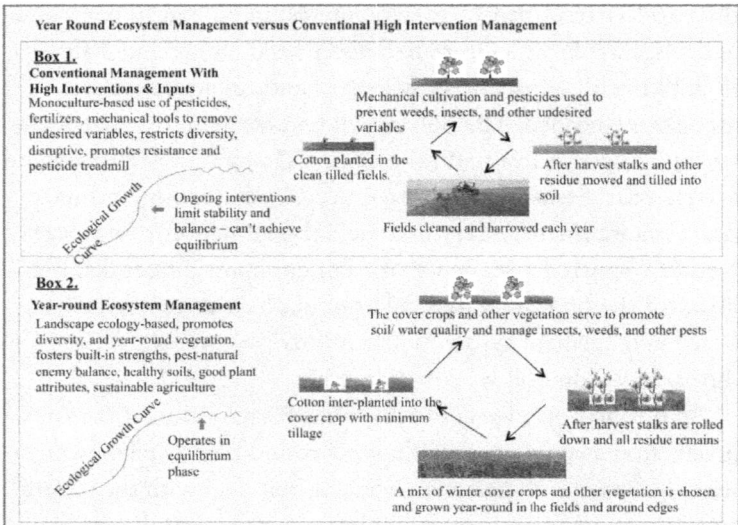

Year Round Ecosystem Management versus Conventional High Intervention Management

Box 1.
Conventional Management With High Interventions & Inputs
Monoculture-based use of pesticides, fertilizers, mechanical tools to remove undesired variables, restricts diversity, disruptive, promotes resistance and pesticide treadmill

Mechanical cultivation and pesticides used to prevent weeds, insects, and other undesired variables

Cotton planted in the clean tilled fields.

After harvest stalks and other residue mowed and tilled into soil

Fields cleaned and harrowed each year

Ongoing interventions limit stability and balance – can't achieve equilibrium

Ecological Growth Curve

Box 2.
Year-round Ecosystem Management
Landscape ecology-based, promotes diversity, and year-round vegetation, fosters built-in strengths, pest-natural enemy balance, healthy soils, good plant attributes, sustainable agriculture

The cover crops and other vegetation serve to promote soil/ water quality and manage insects, weeds, and other pests

Cotton inter-planted into the cover crop with minimum tillage

After harvest stalks are rolled down and all residue remains

A mix of winter cover crops and other vegetation is chosen and grown year-round in the fields and around edges

Operates in equilibrium phase

Ecological Growth Curve

balances into the following season. As one consequence, the pests show up first with a lag time before the natural enemies can be expected.

During the growing season, the crop is kept clean of pests such as weeds, insects, and other undesired variables by thorough cleaning, pre-planting tillage, and other soil preparation and operations, and by diligent mechanical and chemical interventions during the growth and fruiting phase. Use of fertilizers, irrigation, and other inputs are used to ensure a lush, mono-cultural growth of cotton plants from one end of the field to the other. Other plants are considered undesirable and out of place. So, this lush abundance of cotton plants, without alternate vegetation as food sources and shelter for the natural enemies of pests, along with high frequency of mechanical and chemical intervention, creates an environment prone to disruption and resistance, ultimately leading to the pesticide treadmill. This is why, prior to the boll weevil eradication, the number of pesticide treatments for cotton production would sometimes approach twenty per season.

Moreover, the lack of winter cover and the high-intervention approach with substantial removal of the biomass, along with frequent harrowing and tilling, contribute to heavy depletion of organic matter and soil microbial quality, plus extensive water and wind erosion. All of this leads to a host of other issues including lower air and water quality; higher use of fuel, labor, and machinery wear; soil compaction; and the loss of associated wildlife.

Yes, after the boll weevil eradication, we had the opportunity to shift to a less disruptive, environmentally sound, sustainable approach as represented above (Box 2), but it was going to take some time and outreach to bring about such a change in practice. We were up against methods of farming that had dominated pest management in every cropping system for over sixty years. Rachel Carson's call for concern had brought about change, but the change was to move to softer, less toxic pesticides. Still treating the symptoms, in other words. But we had come to understand that the real issue stemmed largely from a lack of understanding of how and why external interventions are disruptive and unsustainable, in contrast with sustainable "built-in" mechanisms, which we had concluded should always be the first line of defense.

I began having discussions about this lack of understanding with Sharad Phatak, a respected pioneer on the subject, and from whom I had gained much insight. We decided to present our case as a profession-wide argument in a highly respected publication. In 1997, he and I, along with Joop van Lenteren and Jim Tumlinson, published a paper in the esteemed journal *Proceedings of the National Academy of Sciences of the United States of America (Proc. Natl. Acad. Sci. USA)*. Our paper, "A Total Systems Approach to Sustainable Pest Management," stressed the urgent necessity for a fundamental shift in how we think about and approach agricultural pest management to resolve escalating economic and environmental problems. We drew on our discoveries to show that an ecosystem is just that—a system, with interactive parts that behaves not like a collection of unrelated pieces, but more like a living organism. We emphasized what we'd learned about the remarkable built-in mechanisms that agricultural ecosystems have, mechanisms that act through a set of feedback loops to maintain balance and to protect against herbivore feeding, diseases, climatic stress, chemical imbalances, and other similar attacks or interventions. To our great satisfaction, the paper turned out to be a major factor in reshaping foundations around sustainable agriculture at grower, research/education, and policy levels. The USDA Sustainable Agriculture and Education Agency adopted the paper for nationwide use as a standard in guiding constituents toward grant proposals, and used it as a standard in developing a sustainable pest management brochure.

The gist of our argument then (as now) centers on the obvious contrast between our sustainable approach making use of the built-in defenses, and the interventionist "treadmill" approach. Figure 9b further illustrates this contrast. The built-in defenses respond only when, where, and at the level needed. They are need-induced and target specific. The chemical SOS signals sent by plants under attack are a perfect example of this. Parasitic wasps searching for these plant feeders, thereby rescuing the plants in distress, create pest control only in fields and around plants with actively feeding populations of caterpillar pests, thus avoiding non-target collateral damage and disruptions.

Figure 9b

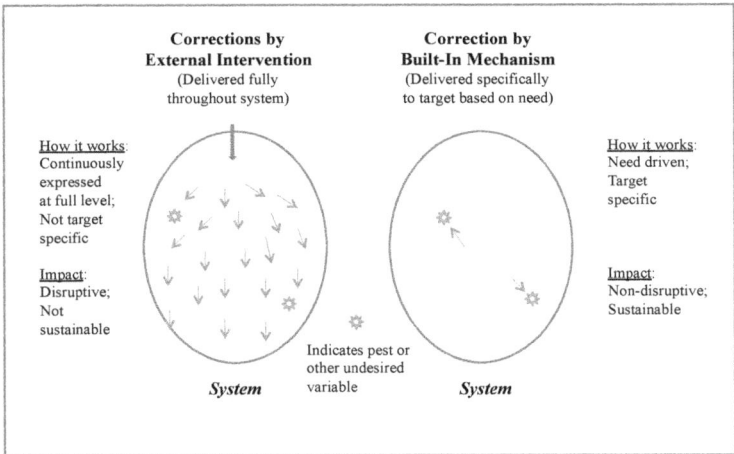

Furthermore, these parasite-host/predator-prey interactions are free of resistance and maintain balance, within fluctuating bounds, through a density-dependent phenomenon, meaning that levels of attack are determined by the availability of hosts or prey. On the other hand, external therapeutic interventions, such as applications of pesticides, act continuously at full level throughout the field without regard to need or target. The consequence is high collateral damage and disruption, and maximum selection for resistance. Next stop: the pesticide treadmill.

The interventionist approach is engrained deeply into not just the agricultural mentality, but in the way we, as a society, think about corrective actions in any system. You can observe the same treadmill effect in how we approach the health of the human body. On the surface, it seems that the proper corrective action for an undesired entity is to apply a direct external counter force, hence a "healthy" dose of antibiotics for infections or painkillers for pain. But there's now a long history in medicine where it can be demonstrated that such interventionist actions never produce sustainable desired effects. They always become less effective requiring more and more to get results. The attempted solution eventually becomes the problem. You can find vivid examples with the growing resistance to antibiotics, and problems of addiction stemming from drugs for treatment of pain or mental distress. Black-market crime is on the rise as people seek illegal sources of drugs, just as it rose during the days of prohibition as an intended solution for alcoholism.

As a matter of fundamental principle, the application of external corrective actions into a system can be effective only for short-term relief. Long-term, sustainable solutions can only be achieved through a shoring up or restructuring of the natural system—in the case of the body, through nutrition, sleep, exercise, etc.—so that natural built-in forces, such as the immune system and other regulators that function on an as-needed basis, act effectively.

The same thing is clear with pest control strategies centered on toxic chemicals and other therapeutic interventions, such as prophylactic treatments. New and "better" pesticides are continually required, just as new and "better" antibiotics are continually required in the field of medicine. It's a constant footrace with nature. The use of pesticides and other treat-the-symptoms approaches are unsustainable and should be the last, rather than the first, line of defense. A pest management strategy should always start with the question, "Why is the pest a pest?" and seek to address underlying weaknesses in ecosystems or agronomic practices that have allowed organisms to reach pest status.

Back in 1990s, with all this in mind, we set out to demonstrate and promote the adoption of pest management in cotton as a part of a year-round landscape management system. This included the use of cover crops. Cover crops had often been used for soil conservation benefits, but their value for enhancing natural enemy/pest balances and relaying them into the cropping system as part of whole-systems pest management and, further, into a year-round agroecology program, had received scant attention. By choosing the appropriate combinations of cover crops with the correct mix of attributes, interplanted with cotton through the use of minimum-till, strip-till, or no-till methods, such an ecosystem pest management approach with cotton production could be piloted.

With the guidance of the University of Georgia Extension Service, and local extension agents, a combination of progressive growers in four Georgia counties were enlisted. By mutually agreed upon guidelines, these growers produced cotton in both conventional and year-round ecosystems on ten- to thirty-acre fields for comparative study and demonstration. The Georgia Cotton Commission and USDA Natural Resources Conservation Service participated in the collaboration as well. Each grower chose the cover crops—wheat, cahava vetch, crimson clover, or crimson clover and rye—depending on their preference.

Philip Haney of my laboratory coordinated the monitoring of the fields and assimilated the data. We knew that by the nature of a single season arrangement, only some of the potential benefits of the year-round ecosystem could be expected to be realized. The benefits would be much stronger if the perennialized system was established over multiple seasons. Yet, the biological and economic benefits in our single-year demonstration were dramatic and encouraging. The densities of nearly every predator group were significantly higher in cotton crops interplanted in the fields with cover crops as compared to the cotton planted in barren clean-tilled fields. In other words, the cover crops were, indeed, relaying natural enemy populations through to the subsequently planted cotton.

Further, the average yield obtained from these year-round management fields was 100 pounds of lint per acre higher than the conventionally managed fields, and the net return after costs was $60 per acre higher. Another benefit for these year-round fields, not calculated in these figures, was that 1.6 fewer per-acre tractor trips, and the accompanying time and fuel costs, were required because of less cultivation and harrowing.

These findings helped expand the acreage of cotton under sustainable practice in Georgia and other areas of the region. Data from the demonstrations were widely distributed by cooperating agencies and stakeholders, including four presentations at the 1997 Beltwide Cotton Conferences. We continued for several years to cooperate directly with local agents in grower group exchanges to advance such practices. And advance we did. Sustainable practices in Coffee County, Georgia, for example, under the seasoned leadership of agent Rick Reed, shot up from 200 acres in 1990 to over 30,000 acres by 2000. One beneficial step utilized was that rotating farmers, including Max Carter, Donnie Smith, and Wayne Fussell would host on-farm "Shade Tree" meetings to discuss their practices and results.

Alton Walker put his money where his mouth was. During all these years, he's pursued the application of these ideas on his own farm. Over the last six years, he's made major progress toward his objectives. His central goals have been to develop a

perennial, vegetation/landscape system that:
- Is largely self-sustaining with minimal input and upkeep requirements;
- Provides for all-around soil and water quality including in the areas of nutrients, moisture, residue, structure, microbial make-up, leaching, wind, and water erosion;
- Provides for good year-round natural enemy/pest balances in all the pest areas of arthropods, diseases, and weeds;
- Provides plants with the right structure, height, and thickness level to accommodate inter-planting of cotton with very limited requirements of mechanical and herbicide interventions.

Alton has advanced all of these objectives. He's established a mix of plants that are largely self-renewing including several clovers, rye grass, rape seed, wild mustard, and others. His soil and water health are rapidly improving. He's able to interplant cotton each spring with a need for minimal strips of herbicide intervention, and with minimal need for mechanical intervention with his own customized equipment. He's had no need to intervene with insecticides in the last three years. He's able to produce his cotton crop with one half the number of tractor trips and less than half the fuel costs as compared to the conventional system. And his yield is averaging one-hundred pounds per acre, beyond the average of the surrounding area.

Alton is now in the process of adding rotations of corn and soybean crops into this system. He has just completed his first year with corn and with excellent results. Historically, reasonable corn production in his location would require irrigation. However, due to the improved water holding capacity and other high quality attributes of the soil, Alton's corn yield averaged ninety-three bushels per acre with only six and one-half inches of rain, and minimal need for other inputs, a very solid yield for the area and conditions. His net profit was approximately $300 per acre.

These real-world practices by Alton show the advantages of the year-round ecosystem management approach. So why, after years of data and demonstration, starting with those early successes in the '90s and continuing to this day on farms like Alton's—why isn't every grower on board with these practices? Good question. For that we need to consider some serious and unfortunate institutional barriers.

Big farming practices have fostered centralization of key aspects of the agricultural industry. Despite certain benefits that emerge from such operations of scale, there are clear downsides. One such downside is a one-size-fits-all mentality that operates throughout the industry. Consequently, as in any field, shifts to practices outside the conventional approach can be challenging. In the case of agriculture, there is difficulty in obtaining the equipment and supplies, along with access to information on methodology. Conversely, a centralized infrastructure can serve to rapidly mobilize a new initiative or technology in a manner that makes it difficult to escape adoption, even in the face of questionable merits.

Over the last twenty-five years, one such initiative in particular has swept through cotton production and other areas of agriculture. In 1996, a spectacular new biotechnology stormed onto the cotton farming scene. That year, having just been cleared for commercial use by the EPA, 1.8 million acres of the genetically modified crop, Bt cotton, was grown in the United States. Genetically engineered with a gene derived from the bacterium, *Bacillus thuringiensis* Berliner. Bt cotton produces a crystal protein toxic to certain pest insects. Subsequently, its use has expanded extensively throughout the world, including China and India. Introduction of this technology reshaped the discussion and greatly complicated the path we were pursuing toward ecologically based cotton production in Georgia and the Southeast.

What is Bt cotton and what makes it so relevant to our discussion? The genes (originally just one, but later up to three genes) used for Bt cotton are coded as toxins for key caterpillar pests, particularly the tobacco budworm, corn earworm, and the pink bollworm. Cultivars with the transgenic presence of these genes continuously produce these toxins at all stages, and by every cell throughout all the plant's tissue as a defense. The use of this technology (primarily in cotton and corn to date, though to a lesser extent for other crops) has been the subject of much debate and controversy. Proponents have argued that Bt cotton is safe for human health and that the toxin is not harmful to other mammals or non-target, beneficial insects. They've argued that a reduced need for pesticide treatments for the targeted pests benefits environmental concerns. Though all parties involved have agreed that there is risk for development of resistance to the toxins, industry and regulatory authorities have pointed to the "high dose"/refuge strategy implemented with commercial use of the technology as appropriate for limiting this concern.

Opponents, on the other hand, have maintained that the widespread, continuous expression of these toxins throughout all tissues of the plants accelerates resistance among insect pests. This resistance will not only neutralize the value of the transgenic plants, but jeopardize our most valuable microbial pesticide, since formulations of *Bacillus thuringiensis* have been a longtime mainstay for organic farmers for a range of pests. They've argued that the technology used in this transgenic manner is premature.

As the arguments continued, the use of Bt cotton expanded on a global basis. By 2010, over 36 million acres of Bt cotton were grown in eleven countries, with the largest acreages grown in the US, China, and India. Though touted by proponents for providing a range of agronomic and environmental benefits, numerous issues continue to surround deployment of this technology. Studies in China, led by Cornell University in cooperation with the Chinese Academy of Science, showed that initial benefits were observed during the first few years but were soon eroded by an increased occurrence of secondary pests, such as plant bugs (Mirids). Consequently, within seven years the Chinese farmers in the study experienced a net income reduction of about eight percent as compared to conventional growers due to the higher costs of the Bt cotton seeds on top of costs associated with these secondary pests.

After its first use in India in 2002, Bt cotton was credited with increased yields and reduced pesticides. However, these benefits were short lived. Recent studies show that proliferation of secondary pests and the development of resistance to Bt cotton by the pink bollworm has led to an actual increase in the need for insecticides. As a result of these expenses, together with the higher seed costs, many farmers in India, rather than gaining benefits in pest management from Bt cotton, are having problems meeting the more capital intensive requirements for producing their crop.

Similar to other parts of the world, the benefits of growing Bt cotton throughout the cotton belt of the U.S. has had mixed results. Since its approval and first use, the adoption rate has expanded from 15 percent of U.S. cotton acreage in 1997 to 37 percent in 2001, and onward to 88 percent in 2020. The extensive use has been met with much acclaim for benefits in yield, the reduction in use of insecticides, and the improvement of environmental concerns. Indeed, the use of Bt cotton in combination with the release of billions of sterile pink bollworm moths over several years helped successfully eradicate the pink bollworm from a limited cotton growing region in the Southwest.

Nevertheless, closer examination reveals significant questions regarding claims of benefits derived from Bt cotton in the U.S. The eradication of the notorious boll weevil was already bringing about major ecological benefits. We've talked about the drastic reduction in Georgia of necessary insecticidal treatments post-eradication (as many as twenty per season down to between three and five), and so it's difficult to ascertain any further net benefits provided by the addition of the Bt technology. Some reports indicate the number of treatments falling even further, down to one or two per season, but that doesn't include the Bt toxin itself, which must be considered a season-long treatment. Though not an external treatment, the Bt toxin is "season-long" treatment applied "inside-out" to every cell of every plant. Furthermore, the Bt cotton seeds are typically coated with a toxic systemic insecticide, predominantly a neonicotinoid, that permeates the plants for some period after germination as a preventive protection against pests of seedling cotton such as thrips, aphids, and other sucking insects. Both these treatments fit the external intervention model shown in Figure 9b, leading inexorably to the treadmill, requiring more and more to get the same result.

In fact, although the Bt toxins have been fully effective in Georgia for the tobacco budworm and generally effective for the cotton bollworm, this latter species has always been less susceptible to the toxin. Further, some "resistance" to the toxin by the latter species has occurred at various locations in the cotton belt, requiring occasional treatment. To combat this "resistance," two new genes have been added (a process called "stacking") and now Bt cotton makes use of a total of three genes, thus the production of three toxins. While it's true that, with the use of these shoring-up steps, the use of Bt cotton has successfully mitigated the budworm/bollworm complex in Georgia and throughout the cotton belt, the success has been dampened by the emergence of several other insects as frequent secondary pests, including aphids, plant bugs, stink bugs, thrips, and white flies.

These elevated occurrences are linked to the adoption of

Bt cotton in multiple ways. Stink bugs have historically not been a pest of cotton in Georgia. But with the deployment of Bt technology, two species, the southern green stink bug and brown stink bug, have become prominent pests. This rise has been attributed by some to the reduced use of synthetic pyrethroids and other broad spectrum insecticides, suggesting that prior to Bt cotton, the use of these insecticides had provided coincidental control of stink bugs. But several studies favor other explanations and reveal how Bt cotton may have led to the emergence of secondary pests as a part of the remarkable web of interplay between plants, herbivores, and natural enemies. We know from our own findings, and from subsequent studies by others, that plants respond to feeding damage by caterpillars and other herbivores by producing a combination of volatile and non-volatile chemicals. We now know that these induced plant responses in turn mediate a range of responses by harmful and beneficial insects vital to the plant's well-being. Not only do the induced chemicals serve to signal natural enemies to their aid, they also serve other roles such as direct repelling, thus deterring egg deposition and slowing the feeding of herbivores. In addition, induced plant chemicals have been shown to provide plant-to-plant communication. The volatile signals alert neighboring plants that the herbivorous pests are in the vicinity, preparing them for a head start with their responses. Several recent studies show evidence stemming from these interactions to explain the emergence of stink bugs and other secondary pests of Bt cotton. These studies suggest that the toxin expression in Bt cotton has reduced caterpillar feeding to the point that the levels of these defensive secondary metabolites are also greatly reduced. Without the deterrence of these chemicals, the stink bugs and other secondary pests, now have open access to the cotton.

If we examine the Bt cotton deployment in the context of Figure 9c, we find that it fits the disruptive reductionist model

Figure 9c

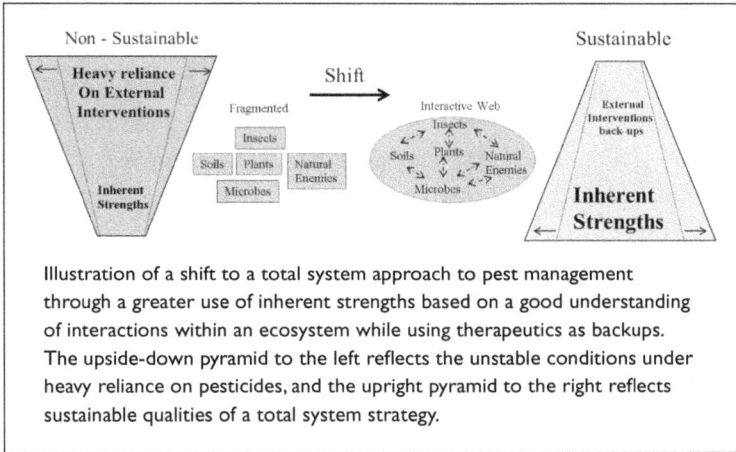

Illustration of a shift to a total system approach to pest management through a greater use of inherent strengths based on a good understanding of interactions within an ecosystem while using therapeutics as backups. The upside-down pyramid to the left reflects the unstable conditions under heavy reliance on pesticides, and the upright pyramid to the right reflects sustainable qualities of a total system strategy.

of the left and is non-sustainable. The toxin is continuously expressed throughout the plant, rather than on an "as needed" target-specific manner as shown in Figure 9b. This approach is, by design, ecologically disruptive and promotes the pesticide treadmill, drawing us away from the opportunity that was afforded us by the historic and remarkable boll weevil eradication accomplishment. Genetic engineering and other such technologies are powerful tools of potential great value in pest management. But if their deployment is to be sustainable, they must be used in conjunction with a solid appreciation of the potential interactions, and in ways that anticipate countermoves within the systems as shown on the right side of Figure 9c. Otherwise, genetic engineering is prone to neutralization by resistance in the same manner as with pesticides.

Centralization—the vertical approach to agriculture, which includes the one-size-fits-all solutions that come from corporate offices far removed from the fields those solutions are supposed to help—is but one institutional barrier that is part of the larger paradigm now at work. If my colleagues and I learned nothing from our experiments with the parasitic wasps, it was the value already inherent in local systems. Our work showed unequivocally how nature works to exploit the resources that are *in place*. The solutions are there if one can only see them, and the solutions are powerful if managed and leveraged properly. The cotton fields we studied were perfectly functioning systems with every piece interrelated. It was a holistic assemblage.

Contrast this natural example with the centralization and specialization that is taking place in agriculture. I'm reminded of cases I've seen where an insect pest specialist in Georgia would confer with an insect pest specialist in California. But just two office doors down, a plant or soil specialist might have more pertinent—more local—information available to that soil specialist. But the opportunity is lost in the fragmented world of the vertical paradigm.

Through the course of all of my work in this field, watching the explosion of big farming, observing the detrimental effects of centralization, studying the harmful consequences of interventionist approaches that ignore the built-in efficiencies of a given ecosystem, I often found myself thinking back to my very first agricultural experiences—those days as a young boy on our simple, little self-contained farm in southwest Mississippi. Those were always wistful thoughts. Indeed, I thought about our little neighborhood, too—the school and the church and the general store. I would consider all that had been lost over the years. And as I would do so, I could not help but wonder if the principles I had learned in my field of study—the discoveries of nature's miraculous web, where everything is connected in a holistic system of built-in efficiencies—could not be applied in a much more general sense. Antibiotics and painkillers provided obvious parallels in the field of medicine, but what about other fields? Indeed, what about society itself? Was the way we approach agriculture symptomatic of a larger societal mentality? If so, what lessons of nature were we ignoring?

Yes, many times, I considered the idea that I was making discoveries that had to do with farming, but also had to do with much, much more.

CHAPTER 10

Community

Somewhere during my time in college, I came upon a quote by Winston Churchill: "We make a living by what we get. We make a life by what we give." The quote stopped me in my tracks like few quotes I'd ever read. Looking back, it's not hard to understand why it resonated with me the way it did. Although I'd never heard the quote before, its truth had been preached to me every day growing up in our little corner of Mississippi. Ours was a caring community where life was about neighbors helping neighbors.

Examples were everywhere. "Swapping work" was a common practice. If you were behind with the plowing of your cotton crop, a neighbor would come over with his mule and he'd plow along with you for a day. You'd return the favor somewhere down the line. It was understood. Maybe you took sick and couldn't finish your harvest before frost. Once word got out, you'd find several of your neighbors in your field taking care of it for you. If a neighbor was laid up with, say, a back injury, you went over with extra firewood, knowing he couldn't chop any himself. Everyone participated this way. Everyone was concerned about, well, everyone. The good people of the community were not just making a living, but making lives.

They were giving back in one form or another to the population of which they were a part. Of course, there was an inherent understanding that this was beneficial for the giver, too. We all had a vested interest in the common good. And nobody else could look after it in the same way we could. Nobody knew our community like we did, the overall wants and needs of the people we worked and socialized with every day. We didn't much benefit from outside help, nor did we seek it.

When I settled into Tifton, Georgia in the fall of 1967 at the age of twenty-four, filled with an abundance of youthful idealism, I brought this attitude, this philosophy, with me. I came with an expectation that I would become a part of the community in which I was now living. I made my living as a research scientist, but I worked hard to balance that with civic involvement.

I knew there was work to be done. The same forces that had driven my father from farming continued to have an ever-expanding impact on community life. The accumulating toll on the economic base and social fabric of agricultural communities was obvious as you drove through the rural areas from Mississippi to Georgia. More and more of the crossroad groceries were idle or vacant. The gatherings of people on store porches in their overalls, chatting or engaged in games of checkers or dominos, were no more. Many of the agricultural-based, one- or two-traffic-light towns were in decline, missing the old bustle of the hardware, feed and seed store, or local grocery market.

I saw these same eroding influences in the Tifton community. But I also saw some vibrancy that appealed to me and I wanted a part in ensuring its future. Little did I know then just how deeply I would become involved and the extent to which the principles behind that involvement would overlap with my professional work.

Beyond agriculture, which was its number one industry, Tifton's economic base was bolstered by a diverse combination of research and education, a budding locally owned medical center, and manufacturing industries, including the manufacture of aluminum products, farm equipment, carpeting, and clothing. The downtown was still in reasonable order with minimal decline, and Interstate 75, with six local exits, had been recently built around the town's western edge, providing fertile potential for a tourist industry of motels and restaurants, although initiatives would be required to prevent this area from harming the downtown.

Shortly after my arrival in 1967, to kick off my involvement, I accompanied Wendell Snow to a Jaycee meeting. The Tifton chapter was strong and active and I was taken by the Jaycee creed, which talked about brotherhood and how "service to humanity is the best work of life." I joined and remained active for several years. In the coming years, I would serve as a deacon in the church I attended, a charter member of the Tift County Historical Society, a coach for several teams in recreational sports leagues (in which my son and daughter both participated), a charter member of the board of directors of the Tifton Museum of Arts and Heritage, a member of the advisory board of the Tift County Training Center for People with Mental Retardation, and a Rotary Club member. Eventually, I would make my way into the political scene of the area, becoming a member and then chairman of the Tift County Board of Elections, elected as a member and Vice-Mayor of Tifton City Council, vice-chairman of the Tifton Downtown Development Authority, member of the Greater Tift Area Planning and Zoning Commission, member of the Regional Board of Mental Health/Mental Retardation/Substance Abuse (for a ten-county area in south-central Georgia), and member of the Suwannee-Satilla Regional Water Planning Council.

None of this is to boast, but rather to underscore the responsibility I felt deep down to immerse myself in the community in which I lived, worked, and raised a family. It was nothing that I hadn't seen in the examples set by the people of my Mississippi community as well as by the people in my new Georgia town. It was common, this level of involvement. Community was strong.

Nevertheless, as time went on, I noticed around me significant and disturbing shifts in my adopted community of Tifton. The forces of centralization and specialization I had seen in farming and small, agricultural-based towns were marching forward at an ever-expanding level. Seemingly every community in America, including the more diverse and vibrant communities like Tifton, were experiencing the influences of these forces. Some of it was good. The internet had made the world a much smaller place, affecting all aspects of our lives, including commerce and travel. Our connectivity encompassed the whole world. In my work I could now exchange documents and in-depth correspondence with colleagues anywhere on the globe in a matter of moments.

But along with this globalized connectivity came a get-big-or-get-out mentality in the world of commerce such as what I had seen in agriculture. Major supercenter shopping malls, chain restaurants, and hotels, all with corporate headquarters in distant locations, began springing up all along the travel corridors. Much of the merchandise was now produced and shipped from other countries with low labor costs. In time, through the magic of the internet, we would be able to order goods from anywhere in the world to be delivered to our front door in a matter of a few days. If we insisted on seeing what we wanted to order, we could still go to the mega chain store that had opened close by. Only later would we recognize the harmful effects—the local bookstore that folded because it couldn't compete with the giant online vendors, or the neighborhood electronics store that went under because everyone started buying their TVs at the supercenters.

Examples of this externalism were soon everywhere, touching every part of our lives. The internet highway was duplicated by vehicle highways that more directly connected our towns and cities, but often by use of bypasses that whisked travelers around once-thriving Main Streets, leaving blighted areas of deterioration. I saw this firsthand in my own town with Interstate 75. Travelers could now skirt around Tifton's downtown, to the detriment of the businesses there. In Tifton and in towns and cities around the country, sprawling subdivisions leapfrogged over mixed-use areas where people once worked, shopped, and lived. Zoning encouraged this with the idea of single usage—residential here, commercial and office over there, retail someplace else.

What it all seemed to lead to was a change in the social fabric of the country. It was a loss of community spirit and sense of belonging. Yes, we were connected to a greater whole in some ways (I could eat at a chain restaurant that served the exact same food in the exact same way as someone in Spokane), but the connection was to something faceless and nameless. People became unmoored from their own neighborhoods in ways that made them unwittingly but inevitably care about them less, paving the way for centralization where larger institutions, both public and private, were now free to come in and help form the conditions under which we went about our daily lives.

We see this in healthcare, where a decision on a patient's treatment might depend on a large and bureaucratic insurance company in another state, or the administrative policies of the mega-corporation that owns the hospital. In education, we see state and federal standards that schools must follow, regardless of local preference or need. We see the loss of local autonomy in human services, social services, and core services like utilities, waste management, and communications.

Although I had already seen this trend in agriculture from the days when big farming had displaced our family farm and put my father out of work, I might have missed it in other areas

but for my work in, of all places, entomology. How strange
that my career studying the habits of insects could have led
me to take notice of such human societal shifts! But wasn't my
community its own ecosystem, just like the cotton field? Wasn't
it at one time beautifully self-contained and self-perpetuating?
Didn't everyone that lived and worked and socialized there play
a part, some larger than others, in how that ecosystem thrived?

We learned the lessons of external interference in the field,
the deleterious effects of interventionism that ignore the built-in
efficiencies of a given ecosystem. And now I was seeing it in society. The truth of the matter is that along with certain efficiencies of centralization that one could argue for, inefficiencies are
just as visible, if not more so. Not long ago, I was discussing
all of this with a friend of mine who lives in Florida. It was December and he shared with me a small but illustrative example
of this loss of local flavor. Shopping for clothes in a large, chain
department store, which will remain nameless here, he was
surprised to come upon a long rack of heavy winter coats. He
mentioned something to the salesperson about the incongruity
of seeing such coats in a Florida store. The outside temperature
that day was a balmy seventy-six. "Yeah," said the salesperson,
shaking his head, "that's what they sent us. They do it every
year at this time. Every store in the nation gets an inventory of
these coats. Last year I think we sold two of them, to people
that happened to be going up north on ski vacations. The rest,
we just always send back in the spring." I couldn't help but
wonder what kind of inventory a locally owned clothing store
would carry in December. I had a feeling it wouldn't have been
heavy winter coats.

This is one rather specific and amusing example of the
larger, more general trend of centralization and specialization
processes by which vital goods and services are provided to local communities. But other examples are everywhere, and often
not as amusing. The providers of these goods and services are

no longer interdependent components of the local community and the end result is a disruption to the balance and, indeed, sustainability of the community, a disruption that can happen in a myriad of ways. A large chain store, such as the one above, selling low-cost clothing could eliminate the market for small local clothing stores, leaving the community with a sole provider. This ultimately reduces the advantages of diversity and local economic self-regulation. In a similar way this situation can occur within the human services profession. When support to people with disabilities, for instance, comes from professional organizations that are not part of the community, the natural community support groups—churches, families, and neighbors—begin to turn their responsibility over to the external agencies.

And this is perhaps the most damaging consequence of external influence. When I was in the Jaycees, we helped set up a training center for people with mental retardation. Then we decided to work with the state for help to grow the program. Great idea in theory, and the state provided much assistance. But I noticed as the program grew that the state took a larger and larger role. Eventually, our little program was essentially swallowed up by the state. We had to step back and before long, decisions were made on our local program without local input. I noticed a degree of indifference growing within the ranks of the locals. What was once an enthusiastic desire to take charge of some given task became a sort of that's-not-my-job mentality. The internal interaction of the system had been effectively removed. The "built-in" advantages were being ignored. An external bureaucracy that ultimately became inefficient had taken over our little ecosystem.

I have seen this dynamic play out in the school system, too. Neighborhoods schools have been consolidated into bigger schools. They're more remote, less accessible, and less integrated into the community. Contrast this with my own schooling.

Our teachers knew our parents. They went to church with them. Saw them at community events. Chatted with them at the general store. If a child was having a problem at home that was interfering with his learning, his teachers knew about it and they could administer the right kind of help.

As mentioned above, administration of the schools is now more "top-down"—under state and federal mandates, more bureaucratic and with less community voice. In response, the communities have stepped back, taking less of a role in the education of their children, adopting a mindset of "it's the professional educators' jobs to educate our students." We send our kids behind the walls of the schools for their education and expect them to come out and get a job. But we know this is not the way people naturally learn. Learning is more of an integrated process involving such things as community partnerships, mentoring, and apprenticeship relations.

These mentor and apprentice relationships are vital processes where learning relates to careers associated with the apprentice's community and "place." And there is an essential need for a teacher who understands this need and knows how to connect on a personal level with the students and their interests, a feature that is reduced in today's larger, more bureaucratic educational system where a high percentage of students do not feel a sense of belonging and consequently don't do well (remember Maslow). What's needed are highly motivated teachers who have a gift for reaching out, relating, and drawing their students in.

My wife Beth is the prime example of such a teacher. Her philosophy? "You cannot feed their minds until you win their hearts." Beth has served as an elementary grade teacher, special education teacher, and high school apprenticeship coordinator, where she worked through the local chamber of commerce and employer organizations to place high school students in apprenticeship positions related to their career interests, including at-risk students. Being named "Teacher of Year" at the third grade, fifth grade, and high school levels, reflects her ability to connect to a broad spectrum of students. In 2012, the Georgia Association of Career and Technical Education presented Beth the Teacher of the Year award in the area of apprenticeship programs, statewide. In this latter role, Beth implemented the SURGE learning model that we developed together (Students Using Resources for Growth and Excellence). Note the summary of this "Learning With a Purpose Model" below.

It's not uncommon that we'll run into someone in Tifton who will recognize Beth from their days as a student in the third grade over forty years ago. Now grown, they'll invariably say something like, "You were my favorite teacher!" I now recognize the look in their eye before they even speak, and can

SURGE
Learning With A Purpose Model

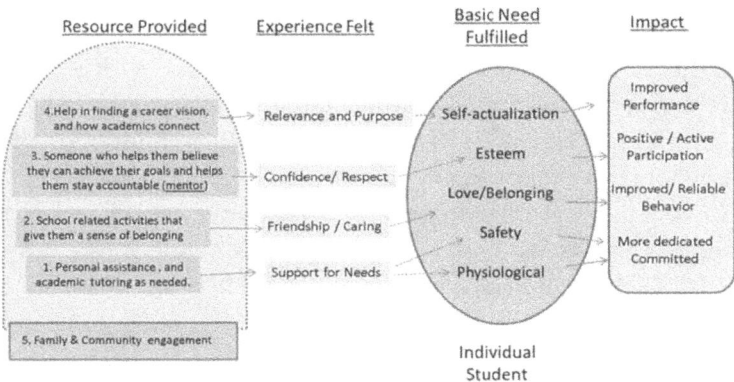

Resource Provided	Experience Felt	Basic Need Fulfilled	Impact
4. Help in finding a career vision, and how academics connect	Relevance and Purpose	Self-actualization	Improved Performance
3. Someone who helps them believe they can achieve their goals and helps them stay accountable (mentor)	Confidence/ Respect	Esteem	Positive / Active Participation
2. School related activities that give them a sense of belonging	Friendship / Caring	Love/Belonging	Improved/ Reliable Behavior
1. Personal assistance , and academic tutoring as needed.	Support for Needs	Safety	More dedicated Committed
5. Family & Community engagement		Physiological	

Individual
Student

The idea of SURGE is to systematically provide students with local resources and relations to deal with personal obstacles, and to provide guidance and support along a pathway to a career of interest, hopefully in the community. The key ingredient is a program where every student feels a sense of belonging and understands how the schoolwork connects to their career goal.

tell that something special happened to them that year. I marvel at the hundreds of former students who are out there that Beth has affected in such a wonderful way. She's gone about her career with the mission of "connecting with students, warming their hearts, and shaping their lives for a positive future." Interestingly, Beth has always admired the success that I enjoyed on an international level in my career. But my impact pales in comparison to hers.

Unfortunately, another weakness of the larger more centralized educational institutions of today is that the pay and benefits are significantly higher at the administrative level. In order for excellent teachers to access these benefits, they must enter the administrative arena, leaving the classroom. But which role is more vital to our central goal? Part of what needs to happen today is that teachers like Beth should be incentivized to remain in roles with direct contact with students rather than being drawn away from classrooms.

Connecting with our communities and feeling that sense of belonging, though critical for our school children, is something that never goes away, no matter how old you become. Today's disassociation and consequent disaffectedness cannot be overstated. When my mother was in her mid-seventies, she had a minor stroke following bypass surgery. Complications led to her being on a ventilator, getting weaker by the day. I had sent her to a world-class hospital in Atlanta, three hours away, and much farther still from my sister, who lived in south Mississippi. I went to see Mother as often as I could. My father came to stay with me and he would visit her, too, but he couldn't stay there alone so he could only visit when I went. Meanwhile, Mother kept getting weaker. Finally, I decided to transfer her back to our local hospital in Tifton where an internist I knew took charge of her treatment. Locally, connected to family and her husband who could now go to spend much of each day with her, my mother began to thrive. Soon she was up and around. My mother would live for ten more years, the beneficiary of the healing power of close human connections.

As I learned in my field, the application of external corrective actions into a system can be effective only for short-term relief. Long-term sustainable solutions have to be achieved through a restructuring of the system so that *intrinsic* forces are allowed to work. We talked earlier about the problems of addiction arising from drugs for treatment of pain and the black-market crime that has accompanied this rise. What we need to address are the causative factors. A possible alternative to tougher laws and failed attempts to limit the availability would be to eliminate the need and market for such substances. This would include an appreciation of the full array of cultural, physical, economic, and social components. In other words, the foundation for such a sustainable approach needs to be built on understanding the attributes of a healthy, "in-balance" community. The development of such communities requires a redirection of the current paradigm and practices so as to maximize the principles drawn from natural systems while moderating the adverse impact of our unfortunate modern trends.

This redirection, to be successful, has to be applied at the local community level. We've all heard the admonition: Think globally, act locally. To bypass the local units of our biosphere and universe, whether human or otherwise, does nothing less than violate our known ecological principles. If you stop and think about it, all the classifications we use to describe components of society—states, regions, nations—are arbitrary groupings. They can serve useful purposes, but they can never substitute for the authentic role of local communities.

In 2000, I published a paper along with Marion Jay entitled "Ecologically-Based Communities: Putting It All Together at the Local Level." We suggested six natural pillars of strength that we believed should be fostered in communities to ensure their health: interdependency, self-sufficiency, self-regulation, self-renewal, efficiency, and diversity/versatility. Conversely, we noted four modern trends that had placed us in undue confrontation with these strengths, threatening the balance and well-being of our communities: specialization/centralization, an interventionist paradigm, high imports and exports, and a therapeutic approach. The figure below summarizes our observations.

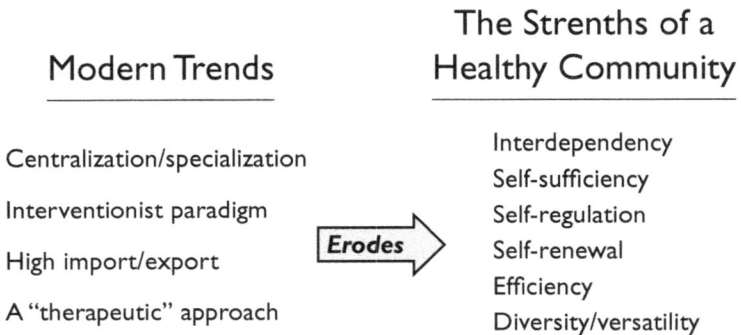

Modern Trends		The Strenths of a Healthy Community
Centralization/specialization		Interdependency
		Self-sufficiency
Interventionist paradigm		Self-regulation
	Erodes ⟹	Self-renewal
High import/export		Efficiency
A "therapeutic" approach		Diversity/versatility

We proposed that protecting the following assets is critical to a healthy community:
- Natural Resources
- Human Resources
- Historical and Cultural Resources

To these, I would also add Economic Resources. Under our modern-day trends these resources are often insufficiently valued and weakly protected.

We were privileged to have Ray Anderson provide us with a foreword for our article. Ray was the CEO of Interface, Inc., one of the world's largest manufacturers of carpet. A year earlier he had written *Mid-Course Correction: Toward a Sustainable Enterprise*, an excellent book, and had become an internationally recognized advocate for sustainability initiatives.

Years before, with a project he termed "Mission Zero," he had made a pledge to eliminate any negative impact his company may have on the environment, a wonderful example of the kind of participation in sustainability that the corporate world can be instrumental in bringing about. Ray loved our ideas about community and sustainability at the local level.

Of course, we were not alone in our ideas. Far from it. In 1992, the Orion Society was founded, a group involved in a variety of societal and environmental issues and publisher of Orion magazine. The Orion Society developed an approach to education called "place-based" education. In place-based education, students and staff alike are engaged in the community, using the community as a major resource for learning. Our SURGE model is an example. It's a means by which to help give students a sense of *place*, the one thing that's most in danger in our modern times. Connecting the classroom with the community is a way to get back to that sense of belonging discussed above, the sense that I had as a young student, where the students and their families and the teachers and the administrators were all a part of the same community.

Around the same time the Orion Society was founded, another group came along with a similar philosophy about moving away from centralization. The Institute for Sustainable Communities was formed to help communities around the world address environmental, economic, and social challenges, mainly with community-based solutions. In 1995, the ISC developed the following elements necessary for community sustainability, which, in 1997, were adopted by the President's Commission of Sustainable Development:

- Leadership, Civic Engagement, and Responsibility
- Ecological Integrity
- Economic Security
- Social Well-Being

These areas are critical to sustainability yet extremely vulnerable to the modern trends of centralization. Think of the historic sections of a downtown being lost to urban sprawl, or the economic centers of a community being moved to distant places, such as our department store example. These are the trends that need to be reversed to avoid the loss of community.

One needs to be careful, however, in speaking in general terms of what will solve any given community's solutions to external threats. To the larger point, specific direction needs to come from within. What's been on the wane in any given community is that community's *identity*, a very specific thing that isn't easily fixed with one-size-fits-all solutions. What governs communities now is the Anytown, USA syndrome, evident almost everywhere. Streets that look alike no matter what town or city you travel to. Homogenous towns that are essentially interchangeable. This is the monoculture farm practice taken to a societal level! The strength inherent in diversity is lost, and with it, the sense of belonging that made each place so unique.

To put it simply, we need to reconnect the individual with the community in which he or she lives. I remain optimistic that this will happen. Organizations like Orion and the Institute for Sustainable Communities have now been joined by others. Movements are underway. I've noticed a trend. People are more interested in shopping local and moving back into downtowns and becoming more involved with their communities. They want food from organic farms and they want to eat at local restaurants instead of the same old chains. It's the (slow) swing of the pendulum. When demand for all of this takes place, market forces dictate that supply will follow and nothing is more powerful (and organic) than a market-driven trend.

Tifton-Tift County has enjoyed considerable success in these efforts and, hopefully, can serve as an example for such a turnaround. Successful actions have included an active "Main Street Program," operating as an arm of the city of Tifton to

market, promote, and coordinate recruitment to downtown of
merchants, businesses, professional offices, and other desired
occupants. The program coordinates a downtown merchants
association, as well as special events and leadership development, and is in partnership with the Downtown Development
Authority to implement programs of low-interest loans, tax
credits, and other incentives for renovation and the preserving
of our historic flavor. For example, a key portion of the residential area of the city was designated a historic district, requiring
permits for external modifications to ensure compatibility
with design guidelines. The purpose was to encourage quality
renovation and the preservation of historic properties, and to
assure downtown property owners that the historic essence of

I spent 16 years as elected member of the City Council and Vice-Mayor of City
of Tifton. I (right) greet a citizen (left) in front of City Hall. A historic hotel in
downtown, now City Hall, serves as an excellent model for adaptive reuse.

the neighborhood would be maintained. The designation also opened access to certain state and federal funds.

Additionally, we were fortunate to have innovative developers who implemented successful "adaptive reuse" of several properties in the downtown area. Two of these developers in particular were Harold Harper, Sr. and Hal Baxley. Of special significance was an initiative by Mr. Harper to lead, through the use of a public/private partnership, the renovation of a large, centrally located hotel to become the city hall on one end and a combination of merchants and other private uses on the other.

Planning, educational, permitting, and incentive programs were developed and implemented in cooperation with the Department of Community Affairs and other agencies to discourage urban sprawl and encourage, instead, "smart growth" and quality land-use patterns that foster water quality and other natural resources, flood prevention, greenspaces/parks/nature, pedestrian-friendly areas, golf cart paths, and quality mixed uses.

The Main Street Program worked with the business leaders along those I-75 exits to advance their interests along with those of the downtown businesses. And it worked at national, state, and regional levels to get satellite offices of state and federal programs located in downtown Tifton, such as the South Georgia office of the Secretary of State, the Georgia Chamber of Commerce, the Department of Labor, and the Department of Veteran Affairs.

All of this has helped significantly in returning Tifton toward the thriving, unique community it once was. And this can happen everywhere. But it can only happen in one way—on the community level, with involved citizens who take an interest in the geographic place they call home, who want to savor and encourage its uniqueness, who want to be an active part of it, and who want to make it happen not just for themselves, but for the people around them, the people they're pleased to call their neighbors.

CHAPTER 11
Family

The year 2001 was a time of changes for our family. In April of that year, Beth and I were married. In July, we moved Mother and Daddy to Tifton. They had agreed that it was time to be closer, and yet they had always been fiercely independent and the move was made with some reluctance and pain.

From that time until their deaths in 2004 (Mother at age eighty-six in February and Daddy at age ninety-three in December), we had many wonderful times at our house, gathered together often around our swimming pool for holidays like the Fourth of July, Labor Day, and various birthdays.

The delight I always felt at these gatherings was mixed with sweetness and sadness. Often, as I would watch my parents playing with their great grandchildren, boundless, wonderful memories rushed through my mind: those early years sitting in Daddy's lap and feeling the security of his weathered farmer hands holding me close; seeing him silhouetted against the late evening or early morning skyline as he walked behind the plow in perfect cadence with our mule; ventures of us feeling our way together through predawn darkness to a turkey hunting position; the delightful meals together (oh, how Mother could cook!), sitting around the table, as we would all gather at the

Mother (Great Grandmother "MaMa Grace") holds Joseph, as Aeriel, Mason, and Austin play around the pool in the background.

home that my parents were finally able to purchase, often followed by late-night card games and laughter with the house filled with children and grandchildren; and, yes, painful early memories of Daddy and Mother struggling, not knowing how they were going to make it on a meager sharecropper's income.

But all the memories were bundled in a context of abiding love and faithfulness. What a blessed life together! Ferd, Grace, and our family—much more extended now—children, grandchildren, and great grandchildren, along with their church family and community neighbors.

If community is important to the well-being of a society, that community starts with the family unit, a microcosm of the larger whole. Like communities, like nature, the family is a system. It's a functioning and interacting group of members

Daddy (Great Grandfather, "PaPa Ferd") holds Mason and Aeriel in his lap in pool house.

that becomes larger than the sum of its parts. The family is nature's way of providing the care and guidance necessary for the species to continue. It's the family that first provides the physiological needs of Maslow, leading to safety, then love, belonging, and esteem. Community, via schools and churches and similar institutions, will ultimately help provide these needs, but the family home is where the individual gets his or her start, for good or for bad.

The best families provide an anchor. I felt this very strongly as a child. We were severely challenged in our material needs, but I scarcely noticed. I felt a sense of safety and belonging. My mother and father were vigilant caretakers. My mother used to say she had eyes in the back of her head, and I do believe that, at times, she did, the way she looked after me and my sister. I have spoken of all that she did for us—the cooking, the clothes washing, the cleaning, all with tools that seem almost primitive to me now. And all the while, being a partner in our farm. To this day, I truthfully do not know how she did it.

There were expectations of us children, too. One frequent admonition, especially as we were getting ready to go somewhere, was, "Remember who you are." This meant to not misbehave, to not in any way sully what it was we were expected to represent, as individuals and as members of our little family. As well, it meant being accountable for our actions, taking ownership of what we did, a value that became one of the strongest in my life. My identity in this regard, my reflection of where I've come from, has and always will, remain an important part of me. It's a cross-generational value. I know I still represent a piece of my mother and father as my children represent a piece of me.

This is a priceless connection of family. I saw it in physical form one day at our local barbershop where four generations of my family have now had their hair cut. The barber who cut my hair, my son's hair, my grandson's hair, and my father's hair had suggested that we all take a picture with him sitting in and around his barber chair. My granddaughter got into the picture, too. This photo, which I treasure, acts as the perfect symbol of that connection, that link from generation to generation. It's a blessing. It's also a responsibility. "Remember who you are."

For our family back in those days of my boyhood, the expectations of us, that sense of responsibility, went hand in hand with the sense of belonging. We were all expected—mother, father, sister, and me—to take part in those activities that were no less than life-sustaining for us. It was a family farm in the truest sense of the meaning. It wasn't just that it was owned by a family, but it was operated by the family with every member lending a needed hand. It was an economic unit, although of course we would not have regarded it that way. And all the pieces that made it function were self-contained. We delegated very little. At supper time, the meal had been provided, in one way or another, by everyone gathered around that table.

The other families in my part of the world functioned likewise and the community became an extension of that familial participation, whether it was to gather together at Uncle Lonnie's store, or worship together in church, or meet at the school to discuss matters of curriculum. This is why participation in my Tifton community became so important to me later on. I found family and community to go hand in hand.

This is a way of life that I was determined to continue, even in the face of the centralization and externalism that was taking over. And I started with that microcosm of community; I started with my family. My son Alan was born in 1968 and my daughter Joanne came along four years later. From the start, it

Shown with my daughter, Joanne, and son, Alan.

My son and family, from left: Mason, Aeriel, Alan, Kathy, Alec, Chance.

was important for me to instill in my children the values that were instilled in me. We didn't have a family farm and my work often took me away from home, but I made it a point to make family time a priority, and to be involved in joint activities starting at an early age.

Both Joanne and Alan were interested in sports. So, I served as a volunteer coach in the recreation leagues, coaching each of their teams for several years, Joanne in softball and basketball, and Alan in baseball, basketball, and football. I found these opportunities to spend such time together with them in the context of their friends to be particularly valuable. When Joanne participated in cross-country during high school, we ran four to five miles together several times weekly as part of her training. Alan and I regularly shared the hobbies of fishing and hunting together. As a family we ate together, we played together, we worshiped together, we worked on projects together, we went

My daughter and her family, from left: Austin, Walt, Joanne, Joseph.

places together, we learned together. We became that whole, the sum greater than the parts.

We all belonged.

As my children grew, they met my fondest expectations in every way. Both were blessings. They have chosen to remain in this community, our joyful times of fellowship are continuing, and they are passing along these same values. Alan and his wife Kathy have three sons, Mason, Chance, Alec, and a daughter, Aeriel. Joanne and her husband, Walt, have two sons, Austin and Joseph. That shared context of love and caring with them provides my life with its deepest meaning and purpose.

With good comes bad. Dianne and I divorced after the kids were out of the house. We had grown apart. The divorce went against everything I believed in and it was painful to experience the sense of failure and loss. In the long run, we both knew the divide was for the best and we made it amicably, our children uppermost in both our minds—the family, never taking a back-seat, even in the midst of a divorce. When Beth came into our family, her side fit right in, like a natural extension. The way it all worked out reminded me of Romans 8: *all things work together for the good of those that love Him according to His purpose.* Beth and I have been married now for twenty years, still spending many good family times together, often around our pool, and with Dianne included.

For years Beth was a highly sought-after swimming instructor. She excelled in that role with the same attributes she possessed as an effective classroom teacher, with students bonding with her and rapidly developing a sense of being safe and confident in her instructions. She taught all four of our grandchildren to swim at very early ages and I watched in amazement as she quickly had them swimming and self-assuredly navigating the deep water. They called her "Mama Kat," picking up on the nickname "Kat" that I use for her. And what "Mama Kat" says goes.

But if the value of family is felt in good times, never is it felt more keenly than in bad times. In 2006, we lost Beth's son in an automobile accident. They were close and I had become close to Jay, too. He was just shy of twenty-eight. Nothing, of course, prepares one for the pain of a lost child, but if there is any place where comfort may be found, that place is in the bosom of the family. "I can never replace Jay," my son told Beth, "but I want you to know that I'm going to be the best son to you that I can possibly be." The need for belonging, the need for safety and security in a seemingly arbitrary world of sometimes sudden change, is a constant, no matter how far up on Maslow's famous pyramid one scales.

Jay was a kind and caring young man with immense and lasting influence on his circle of friends. The power of that influence lives on. Several of Jay's friends moved quickly to establish a Jay Deason Fund for Children, to celebrate Jay's memory,

Beth and her son, Jay.

and to honor the fact that he was completing his degree in early childhood education. He was just getting ready to embark on a career as an elementary grade teacher at the time of his death. For fourteen years now, this fund has been remarkably success-ful, providing thousands of dollars in funding for the needs of children throughout the Tifton/Tift County community. This lasting impact of Jay's life has been of extraordinary value to Beth and me.

Family is vital, but an idea that is becoming lost is the extension of family into the surrounding neighborhood. But it needn't be so. All it takes is a little effort to get to know the people who live nearby. People often live right next door to each other yet remain strangers for years, but it doesn't have to be like this. Beth and I have made it a point to get to know our neighbors, even to host get-togethers. We have come, as a neighborhood, to look out for each other, to care for one another. And if technology has allowed us as a society to drift apart, it has also provided ways for us to come together. As just one example, neighborhood message boards are now available online where you can interact with the family down the street from the comfort of your own home. There are ways to connect, even in the most seemingly disconnected of times.

Today's disconnectedness isn't merely technological, of course. It's also hugely geographic. We move around a lot. It's becoming less and less common to have more than one generation living in any given area. Sometimes the divide is as far as the country is wide. Sometimes it crosses countries.

We've moved away from something else besides family, too. We've moved away from nature. Our environments are far more urban. We no longer understand the language of nature. Our kids have not only stopped playing in the woods and fields, they've stopped playing outside. Today's children spend, on average, more than three hours a day on some form of electronic media, leading to what Richard Louv, author of *Last Child in the Woods*, calls Nature-Deficit Disorder.

The family is the one institution that can reverse these trends. The more time parents spend with their children—playing together, working together, learning together, worshipping together—the closer become the bonds, the closer become the strength and power of family. And if it starts with the family, the family starts with the parents. They are the first teachers and they set their examples, for good or for bad, by their actions. As a poet once said, "No written word, no spoken plea can teach our youth what they should be. Nor all the books on all the shelves, it's what the teachers are themselves."

Teaching by example has to include time spent outdoors. We must, all of us, relearn nature's language. Of all the things that have been lost to modern trends, this is perhaps the most disturbing and, potentially, the most dangerous. Disconnected from nature, we as a society, no longer respect her ways and if we do so, we do so at our peril. We believe we can master her with our external interventions, but ever since Rachel Carson's Silent Spring, the alarm has been sounding. It is not too late to answer.

Back to nature, back to the community, back to family. Or, perhaps, I should say, forward to all of these things. Note well their intersecting quality. We are all family, in a larger sense, and all connected to nature whether we realize it or not. We all belong, and never is this more crucial than in that wonderful, natural, institution of the family, anchoring as it does the larger institution of community, where we can find our place and our comfort in fellowship. "In the sweetness of friendship," said the poet-philosopher Khalil Gibran, "let there be laughter and sharing of pleasures. For in the dew of little things, the heart finds its morning and is refreshed."

CHAPTER 12

Builders for Eternity

*"Will the LORD be pleased with thousands of rams, or with
ten thousand of rivers of oil? ...He hath shewed thee, O man,
what is good; and what doth the LORD require of thee, but to
do justly, to love mercy, and to walk humbly with thy God?"*

Micah 6:7-8

In September of 2010, I made a return trip to Lincoln County, Mississippi to attend the fiftieth reunion of my high school
graduation class. Beth and I took the opportunity to visit the
site of Uncle Lonnie's grocery store.

The store and nearby house were gone, as were other buildings I remembered in the area. Uncle Lonnie's store, the symbol
and, indeed, essence of community, of sense of place, was no
more. Beth and I walked around the area for nearly an hour as
I described how things had looked and the sounds and activities
that had taken place. In my mind, I could still see and hear it
all. There was the store, the steps up to the porch, the people
coming and going. This person always came this way and that
person from that way. Some drove, others walked down this

road or came along that path. I remembered who they were, these neighbors and friends. I could see them sitting and standing here and there about the store and on the porch. I recalled as if it were yesterday how they looked and sounded. I could hear the teasing and the laughter and the fellowship.

As younger children we played games close by. In the evenings, we chased lightning bugs. I remembered frequently climbing a tree that was no longer there. I pointed out to Beth the fields where, as older kids, we played baseball and basketball. The fields grew up just as we did. They were filled with pine trees now. I recalled coming back to the store to drink soda pop on the porch after those long games in the hot summer sun. My preference was Royal Crown Cola because it came in a larger bottle, and I would drink it slower and slower as I got near the bottom, swirling it around before taking that last swallow. Soda pops were a nickel and I still remember the day Uncle Lonnie raised the price to six cents.

Beth and I drove around the area and I noticed that not only was the store gone, but the overall community had declined. Within one area, where I had remembered twenty homes, there were now five. And even with those five, the front porches were empty and there were no children playing in the yards.

This place, like so many others throughout the rural areas of Mississippi, and the nation for that matter, represented the essence of America at its best, where American dreams, including mine, had been born. Our forefathers and foremothers had built those communities over several generations. They had faced the horrors of two world wars and a depression to protect them. They had worked, fought, and prayed for a better tomorrow for their children and their children's children. I had soaked up their hopes and desires on those grounds, and had sought to pass everything on a step better for my children.

Returning back to Tifton, I couldn't shake the contrasting visions of my childhood home, the one I saw with Beth, and the one I'd grown up with—the one that had been crafted so lovingly from those who had come before me. The thoughts I had reminded me of a poem by R. Lee Sharpe:

Isn't it strange how princes and kings,
and clowns that caper in sawdust rings,
and common people, like you and me,
are builders for eternity?

Each is given a list of rules;
a shapeless mass; a bag of tools.
And each must fashion, ere life is flown,
A stumbling block, or a stepping-stone.

And I began to wonder: was my generation building stepping stones or had things gone awry and were we now beginning to leave stumbling blocks? Putting the question another way, what happened—what *really* happened—to Uncle Lonnie's store?

I had been keenly aware, for some time, of the store's closure, as well as the dwindling of the surrounding community. The "Get Big or Get Out" movement for farming had arrived there in the 1950s, and my family experienced its ultimate sting with the necessary departure of my father from farming during my high school years. Naturally, I was troubled by this deterioration of my beloved community and its way of life. But far more disturbing was how the `disappearance of Uncle Lonnie's store and the fading of its surrounding community were symbolic of a fundamental weakness that threatens the sustainability not only of rural communities, but our civilization in general at regional, national, and global levels.

In brief, this weakness is the collective impact of the tools, practices, and mindsets created as an unintentional consequence of the industrial revolutions. Since the late eighteenth century, we've see innovations and developments that include the steam engine, internal combustion engine, electricity, age of science, mass production, transportation technology, telephones, computers and digital technology, and global connectivity. With all of this has come urbanization, centralization, reductionist approaches with associated specialization, global travel and communication. These changes were slow at first, and then snowballed in the past half-century. Business shifted from horizontal and integrated interactions at the local levels to vertical, centralized/specialized organizations with remote headquarters and multiple chain locations. These organizational designs promoted the movement of people and decision-making to urban centers with corresponding reductions of people from the rural communities. At the same time, the introductions of large automated tools such as tractors, and chemical fertilizers and pesticides and other advances, further reduced the need for human labor and encouraged the interventionist paradigm and therapeutic management approaches. Finally, all these innovations and associated practices brought ever increasing demands for energy and natural resources while adding to our waste management and pollution issues.

These innovations and practices have become an integral part of every aspect of our society. For agriculture in particular, I had something of a front-row seat to these changes from the middle part of the last century forward. I grew up in a very modest agricultural environment that was absent modern conveniences and then spent my adult life as a scientist studying insects in particular, but nature in general. Perhaps, then, I have apprehended in a more personal way our inextricable link to nature and our dependence on it. None of our innovations can erase this link or dependence. And nowhere is this more apparent than at the level in which we feed ourselves and sustain our lives. Agriculture is the foundation upon which all else rests and should, therefore, be the model for working in harmony with nature. As goes agriculture, so goes the rest of the world.

Where we live, our biosphere, powered by the sun, is no more than a thin shell around the earth, varying roughly between five miles below sea level and five miles above. It is from this area that we acquire everything we need to survive on this planet. Two fundamental activities keep this biosphere in existence: the transference of energy and the recycling of materials. For millennia, people lived in a way that maintained these activities as a unified, sustainable cycle. Societies worked with what they had, generally by necessity, never taking more than what their little portion of the earth could give, nor disrupting the order and balance of the cycle. You either did that, or you risked eventually dying out. This reality has not changed.

Working with what one has and not taking more than what can be given is the way we lived during my early years in Mississippi. It was a way of life that was highly integrated with nature, all the way down to structuring our days to take advantage of the daily cycle of the sun. We were not just connected to nature; we were embedded in it. Our neighbors were, as well, and so we were interwoven with the larger community. We all saw ourselves as a team, sharing and jointly protecting what was in the interest of the common good, within the time of our generation and future generations. Rarely did you see posted land, nor did you see a lack of respect for mutual interests. We were horizontally organized as a self-contained community under the classic "butcher, baker, candlestick maker" model, with only minimal needs from the outside. Our farming was multifunctional and diversified, simple though it was, with combinations of domesticated livestock and plants on the farm. We had plentiful resources available for hunting, fishing, and gathering. We understood the connection between the land and the food on our table.

For me, as a youngster with a robust curiosity, it was a playground of endless adventures, abundant with insects, birds, plants, and other wildlife. None of which were threatened with extinction. Surely, it was these experiences that drove me to go on in later years to a scientific career learning even more about the remarkable aspects of the plants and animals.

But when the innovations of the 1950s arrived, the paradigm began its shift. The era of big machinery and fossil fuels and urbanization and high connectivity spurred the trends mentioned earlier of centralization and interventionism. We went, as a society, from being embedded in nature to trying to control and dominate it. We lost sight of the truth in an oft-quoted statement attributed to Chief Seattle: "Man did not weave the web of life, he is merely a strand in it. Whatever he does to the web, he does to himself." Our little farm was its own ecosystem, but one that had to give way to the now larger system that swallowed it up, with tractors as big as the shed in which Daddy kept his plow, acres of single crops on heavily tilled land with high inputs of fertilizers and pesticides brought in from elsewhere, and decisions made in office buildings several states away.

With this dynamic came the ineluctable loss of community and all that followed. We had been connected not just to the land, but to the place where we lived. Life had been about relationships, with nature and with each other. We were impoverished, but we wanted for nothing. But the massive changes engendered the idea that everything can be accomplished better on a mass scale and all problems have universal, one-size-fits-all solutions. Big box stores, highway bypasses, medical decisions made by large insurance companies, manufacturing outsourced to other countries—all of this came about by the forces outlined earlier: specialization/centralization, the interventionist paradigm, high imports and exports, and therapeutic approaches. And this held true whether it was manufacturing, medicine, retail, education, or whatever the institution. In each case, the local system became disrupted, no less so than the cotton field's perfectly orchestrated system becoming disrupted by man-made chemicals, fertilizers, monocultural planting practices from ditch row to ditch row, and tilling that strips a field back to zero each year.

When systems get disrupted from the outside, they inevitably break down. In the best case, they simply do not work as efficiently. At least not for long. No rational person would argue against bringing in a man-made pesticide to save a crop that was in mortal danger. But such a solution needs to be targeted and it needs to be used taking into account the needs of the local system. It needs to be used in a complementary way, in other words, and never as a substitute solution. As this book is going to press, we are witnessing a pandemic that is challenging our food and supply chains, built as they were, in our age of centralization. An average pound of food travels 1,500 miles before reaching the table. This centralized, vertically designed model, as opposed to a more holistic and natural community-by-community system, places us at high food-security risk. In many cases, we are now seeing how the closing or slowing of a distribution center in one part of the world materially affects populations elsewhere in the world. Globalization, partially responsible for the pandemic's spread in the first place, was not built to handle the impact of it, making us vulnerable at the local level, the place in which we ought to be our strongest.

Agriculture has led the way for the sea of change we have witnessed over the last century because systems are natural mechanisms and no industry is closer to nature than agriculture. When rural farming became disrupted, we began to mess with nature herself at her most fundamental level. What we did, to put it another way, is lose our connection to nature. We did this most obviously with our new farming techniques, and, collectively, we did this almost unconsciously with our move toward centralization and urbanization. Every step removed from the family farm was another step removed from nature.

The move towards "bigness" has impacted our resources and our reserves. The sustainable model of the local system has been lost to a model that takes from the earth in a way that exploits earth's reserves, mining resources in a single lifetime that took thousands and thousands of years to accumulate. In this way, we have not only lost our connection with our planet, we are working at cross-purposes with it, taking for ourselves at the cost of future generations. Environmentalists talk these days of the three R's: reduce, reuse, and recycle. This is a fine approach, so far as it goes, but what's ultimately needed is a way back to regeneration and renewal. If you're walking down the wrong path, slowing down might seem helpful, but you're still walking down the wrong path. In the long run, nothing less than a return to self-sustaining systems will be able to get us back to the way of our forefathers and foremothers who took what the earth could give and no more.

So how do we get back? Where do we even begin?

Back in the 1970s, in the field of environmentalism, this formula first appeared: $I = P \times A \times T$, [13] where "I" stands for ecological *impact* and is a function of the interdependent forces of *population*, *affluence*, and *technology*. Notice that an increase in technology (like population and affluence) provides a multiplicative effect on the impact of the environment. But it needn't be like this. Ray Anderson proposed a new equation. To him, "T" represented the technologies that came along since the industrial revolution, the ones we saw in such force through the twentieth century, the ones that took, and continue to take, from the earth in unsustainable ways. Anderson called these "T_1" technologies and imagined, instead, new, sustainability-focused technologies, which he called "T_2." His equation looks like this: $I = P \times A/T_2$. Note the difference. Now, the addition of technology (increasing the formula's denominator) serves to

---

13   Generally attributed to Paul and Anne Ehrlich.

decrease the net impact.

Let's examine this line of thinking as applied to agriculture, using the T_1 versus T_2 concept only, placing the P & A aside, and using Figure 12 as a guide. The upside down pyramid on the left represents conventional modern agriculture with emphasis primarily on external interventions (T_1 technologies – e.g., pesticides, clean tillage, synthetic fertilizers, etc.) and little consideration for the inherent strengths and regulators. As indicated, these disruptive interventions continue to weaken the inherent strengths with a treadmill of negative impacts. On the other hand, a shift to emphasis on inherent strengths or use of T_2 technologies is indicated on the right. These "built-in" re-newable technologies work in harmony with the natural system and can be represented as a value in the denominator, whereby their use moves the negative impacts toward zero.

We can apply this same model to other systems is a similar

Figure 12

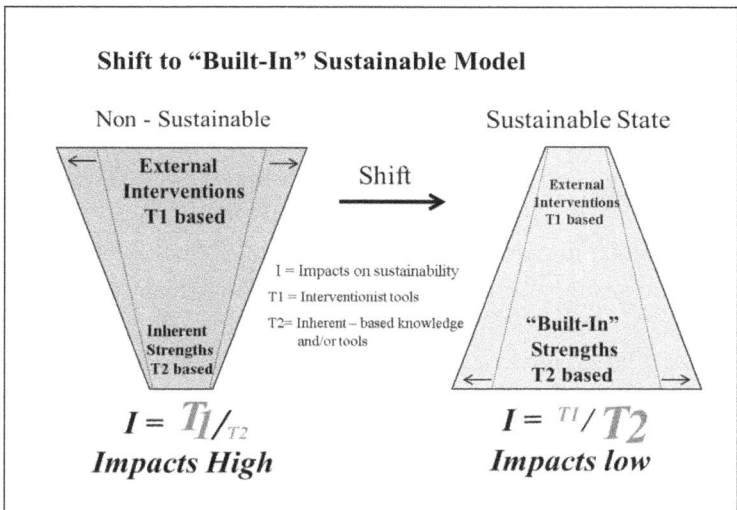

Shift to "Built-In" Sustainable Model

Non - Sustainable   Sustainable State

External Interventions T1 based   Shift   External Interventions T1 based

I = Impacts on sustainability
T1 = Interventionist tools
T2 = Inherent – based knowledge and/or tools

Inherent Strengths T2 based   "Built-In" Strengths T2 based

$$I = T_1/_{T_2}$$
Impacts High

$$I = {}^{T_1}/T_2$$
Impacts low

manner. For example, in human health management where the external interventions of T_1 represent antibiotics, pain killers, and so forth, and the inherent strengths of T_2 represent nutrition, rest, exercise, and the like. In the case of human communities and/or components of communities, the model would apply with the external interventions of T_1 representing large cookie-cutter, centralized programs versus T_2 programs like place-based education that emphasizes connections of classrooms and the local communities.

Technology itself can be a way back. Farming practices need to go in the direction outlined earlier, taking advantage of built-in mechanisms for sustainable agriculture. In other words, technology itself is not the problem, rather the wrong technology or technology applied in the wrong way is the problem. Conversely, the right technology (T_2) can provide the solution.

In the case of agriculture, I don't have to search too hard to produce a specific example of the T_2 technology. Our discovery of the remarkable ability of plants, when fed on by caterpillars, to emit "SOS" signals that "call" parasitic wasps to come to their aid, and the wasps learning and understanding these calls in very specific ways, is a perfect case in point. This is a type of technology that acts in a target-specific manner and on an as-needed basis. It's sustainable and not disruptive. And nature is full of such holistic systems with built-in solutions. No outside intervention needed. These built-in tools—T_2 innovations—are just waiting to be discovered if we would but seek to understand and deploy them instead of falling back to disruptive, non-sustainable interventions. Alton Walker's 600 acres need to become the norm, not the exception.

John Ikerd, professor emeritus of agricultural economics at the University of Missouri and author of *Sustainable Capitalism, A Return to Common Sense*, and *Small Farms are Real Farms*, talks about the concept of agroecology—a "unifying framework" for agriculture and ecology, recognizing that everything is connected, but further recognizing the "ecology of place," the unique nature of each ecosystem. It's a framework that also incorporates economics and sociology. Economics alone can't take us to where we need to go because, as Ikerd points out, from a strictly economic standpoint—without the promise of a return, in other words—there's no reason to care about others, to care about our grandchildren, or to care, for that matter, about nature. Reiterating the idea that sustainable agriculture, taking advantage of the built-in mechanisms of efficiency, is where we need to return, Ikerd reminds us that there "is no better place to reconnect purpose with people and place than in a local, community-based food system."[14]

What's most needed to implement Anderson's T2 technologies and to follow Ikerd's path of agroecology is a change in mindset. Sustainability is an interest and a concern for, above all else, future generations. This concern is what needs to be uppermost in our thinking, in our practices, in our policies, and in our hearts. As Ikerd puts it, "We truly care about specific people and places only if we believe it is our responsibility, our purpose, to somehow contribute to the goodness of those people and places. It is this ability to contribute to the goodness or betterment of humanity and of the earth that gives meaning to our day to day lives."

This will require, above all else, a reconnection with nature. If I am saying nothing else here, I am saying that our disconnect with nature—her beauty, her power, her amazing ability to give—is, more than anything, the greatest threat to our survival.

The real beauty is that we don't need to go very far or acquire any special skills to discover or rekindle our connection. I can say this based on my own personal experience. I grew up with nature. But my appreciation grew exponentially through

---

14 "Saving the World by Reconnecting People and Place," prepared for presentation at the Tennessee Local Food Summit, Montgomery Bell Academy, Nashville, TN, November 29-December 2, 2018.

my professional work, where I was able to apprehend nature in a different, more complete, way, making her even more amazing to my eyes. My life journey opened up vistas I had never seen nor could ever have imagined. The funny thing, however, is that what I found had been there all along, right in front of me.

It's right in front of all of us, no matter how far we go or how technologically advanced we become. Nature, in all of her self-sustaining glory, is still here, waiting for us, with solutions already built in, if we could only see them. "The real voyage of discovery," wrote Marcel Proust, "consists not in seeking new landscapes, but in having new eyes." With new eyes and a new language we can create a life of wonder for future generations. We can fashion stepping stones. We can be builders for eternity.

Where They Are Now

A list of participants of the Team Program reflected in the Chapter Eight portion of the book, showing their native country (in parenthesis) and where they are currently. A few are not included only because the information could not be obtained at the time. All of the participants were important and valued.

Richard L. Jones, (USA), Retired, Former Dean for Research and Director of the Florida Agricultural Experiment Station, Gainesville, Florida

Jim Tumlinson, (USA) Ralph O. Mumma Professor of Entomology, Co-director, Center for Chemical Ecology, Pennsylvania State University

Louise Vet, (The Netherlands), former Director of the Netherlands Institute of Ecology, The Netherlands

Joop van Lenteren, (The Netherlands), Professor Emeritus Professor of Entomology, Wageningen University, The Netherlands

Consuelo M. De Moraes, (Brazil) Professor, Biocommunication, ETH Zürich, Zürich, Switzerland

Felix Wackers, (The Netherlands) Director, Research and Development, Biobest,Group, Professor Plant-Insect Interactions, Lancaster University

Keiji Takasu, (Japan) Professor, Faculty of Agriculture, Kyushu University, Japan

Ted Turlings, (The Netherlands) Professor, University of Neuchâtel, Switzerland, Director of the Centre of Competence in Chemical Ecology

Michael Keller, (USA) Professor Emeritus, School of Agriculture, Food and Wine, University of Adelaide, Australia

John Ruberson, (USA), Professor and Head, Department of Entomology, University of Nebraska - Lincoln

Anne Marie Cortesero, (France) Professor, Institute of Genetics, Environment and Protection of Plants, University of Rennes, France

Oscar Stapel, (The Netherlands), Employed with a private company providing landscaping and related services in the Rennes, France area. Note: he and Anne Marie met while at Tifton and were later married and live in the Rennes area.

Lucas Noldus, (The Netherlands), Founder and CEO, Noldus Information Technology, Wageningen, the Netherlands. (www.Noldus.com)

Franck Herard, (France), Entomologist, European, Biological Control, Laboratory, USDA-AR S, Montpellier, France

Hans Alborn, (Sweden), Research Chemist, Center for Medical, Agricultural and Veterinary Entomology, Gainesville, Florida

Dawn Olson, (USA), Research Entomologist, USDA-ARS, Tifton Campus, University of Georgia, Tifton, Georgia

Paul Pare, (USA), Professor of Chemistry & Biochemistry, Texas Tech University, Lubbock, Texas

Fred Eller, (USA), Research Chemist, Functional Foods Research Unit, National Center for Agricultural Utilization Research, USDA-ARS, Peoria, Illinois

Philip McCall, (United Kingdom), Liverpool School of Tropical Medicine, Pembroke Place, Liverpool UK

Torsten Meiner, (Germany), Institute for Ecological Chemistry, Plant Analysis and Stored Product Protection, Federal Research Centre for Cultivated Plants, Julius Kühn Institute, Berlin, Germany

Ursula Röse, (Germany), Associate Professor, University of New England, Biddeford, Maine

Yvonne Drost, (The Netherlands), Epidemiology and Research Support, Triimbost-Instituut, Utrecht, Netherlands

Marco D'Alessandro, (Switzerland), the Swiss Environment Ministery BAFU, Switzerland

Naoki Mori, (Japan), Professor Chemical Ecology, Kyoto University

Claire Bonifay, (Switzerland), Biology Teacher, Kantonschule Wettingen, Switzerland

Moukaram Tertuliano, (Benin), Post-Doctoral Fellow, University of Georgia, Tifton, Georgia

Jeff Tomberlin, (USA), Professor, Department of Entomology, Texas A&M University, College Station, Texas

Douglas Whitman, (USA), Professor of Biology and Curator of Insects, Illinois State University, Normal, Illinois

Index of Common Terms

More Books from Acres U.S.A.

Order by calling 1-800-355-5313, or online at www.acresusa.com. Also, enjoy our free informational articles from many of our popular authors at www.ecofarmingdaily.com.

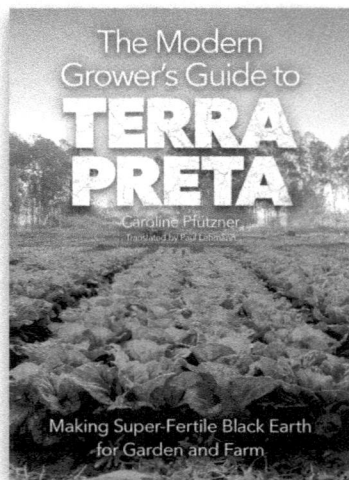

Terra Preta
by Caroline Pfützner

Translated into English for the first time, author Caroline Pfützner introduces us to terra preta, or black earth of the Amazon, what is considered the most fertile soil in the world. Rightly so, because this ultra-rich, living material literally builds a permanent humus layer on the land. The true results of working with this almost miraculous substance are healthy plants and a rich harvest — without outside fertilizer inputs.

And even better, widespread use of terra preta would actively protect the climate. This practical book by a world authority on the subject—available in English for the first time—practically guarantees success in production and application of terra preta whether in the garden, raised beds, larger growing operations, or simple balcony boxes. Practical examples from commercial-scale agriculture illustrate the true potential of terra preta.

Making black earth by yourself. Learn step-by-step how to make top-quality terra preta yourself.

Using terra preta in your garden or farm. See how the author grows healthy, bountiful crops organically without synthetic fertilizer.

Practical in the extreme. Far from a cumbersome scientific text, benefit from the rich, practical experience of the author.

Extensive background knowledge. Understand the deeper story—historical and scientific—of terra preta and its implications for modern times.

#7566 • Softcover • 176 pages • Copyright 2019 • $28.00

The Farm as Ecosystem
by Jerry Brunetti

In The Farm as Ecosystem, natural product formulator and farm consultant Jerry Brunetti brings together a wealth of education and uncanny observations in this probing volume on the interconnected dynamics of the farm — geology, biology, and diversity of life. Learn to look at — and manage — your farm very differently by gaining a deeper understanding of the complementary roles of all facets of your farm.

With this unique perspective, the author guides the reader on a journey through the modern farm as an ecosystem, providing intimate anecdotes and comprehensive details that appreciate all dimensions of the farm. Brunetti's work is invaluable to the contemporary farmer and to those seeking an original appraisal of farming and its future.

Topics covered by The Farm as Ecosystem include:
- The physical, chemical and biological aspects of soil;
- Understanding compost and compost tea;
- Working with foliar nutrition;
- The roles of trace elements in farming;
- Water and your farm;
- Cover cropping systems . . . and more.

The book eco-farmers everywhere have been waiting for is here.

Copyright 2013 • Softcover • 352 pages • $30.00

About Jerry Brunetti
JERRY BRUNETTI, 1950-2014, worked as a soil and crop consultant, primarily for livestock farms and ranches, and improved crop quality and livestock performance and health on certified organic farms. In 1979, he founded Agri-Dynamics Inc., and confounded Earthworks in 1990. He spoke widely on the topics of human, animal and farm health

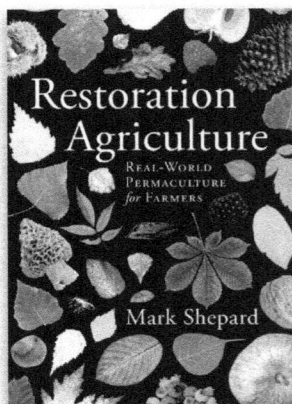

Restoration Agriculture
by Mark Shepard

Around the globe most people get their calories from "annual" agriculture—plants that grow fast for one season, produce lots of seeds, then die. Every single human society that has relied on annual crops for staple foods has collapsed. Restoration Agriculture explains how we can have all of the benefits of natural, perennial ecosystems and create agricultural systems that imitate nature in form and function while still providing for our food, building, fuel and many other needs—in your own backyard, farm or ranch. This book, based on real-world practices, presents an alternative to the agriculture system of eradication and offers exciting hope for our future.

Copyright 2013 • Softcover • 339 pages • $30.00

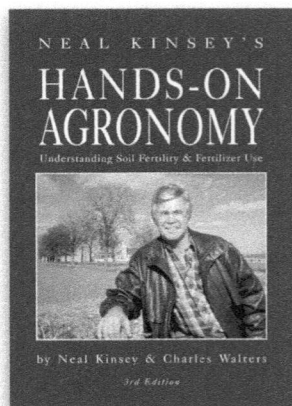

Hands-On Agronomy
by Neal Kinsey

The soil is more than just a substrate that anchors crops in place. An ecologically balanced soil system is essential for maintaining healthy crops. Hands-On Agronomy is a comprehensive manual on effective soil fertility management providing many on-farm examples to illustrate the various principles and how to use them. The function of micronutrients, earthworms, soil drainage, tilth, soil structure, and organic matter is explained in thorough detail.

Neal Kinsey demonstrates that working with the soil produces healthier crops with a higher yield. To that end, he provides an understanding of eco-agriculture as a viable enterprise that is both naturally and commercially sustainable. This work offers advice that is both highly valuable and applicable to the farmer of any level and promotes an ecological understanding of the farm from the ground up.

Copyright 2013,1993 • Softcover • 391 pages • $35.00

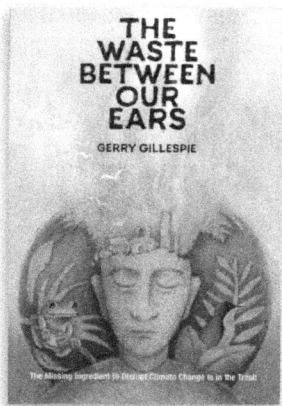

The Waste Between Our Ears
By Gerry Gillespie

We know the dire situation:

- Due to industrial agriculture and the overuse of chemicals, we are actively degrading soils to the point that only a few decades of farming seasons are left;
- The World Wildlife Fund reported that our way of living has wiped out 60 percent of all mammals, birds, fish, and reptiles since 1970; and
- Chemical destruction of biological processes poisons our soils, our air, and our water.

The time for effective change is shrinking.

In this book, writer, researcher and advocate Gerry Gillespie outlines how we can create a global solution, and it starts between our ears. In order to restore our world ecosystems and our vital soils, he wants to change how we think about our trash. Readers will learn why we all need to change our mind about waste management systems, how to reconnect our organic waste to local soil and food growers, and why this leads to more local jobs.

With chapters about source separation, soil management and climate change, and practical approaches to zero waste, Gillespie presents a practical, logical argument for one way to save the world and grow a local economy. In economic and environmental terms, clearly explains how, if waste were collected as source-separated products, more than half of it could be returned to soils as quality compost and biological products. A very large percentage of the remainder can be put back through recycling, re-manufacturing and reusing. And it's not just theoretical. Gillespie details how this reduction of waste is already being achieved in parts of the world and how we could do it globally ... if we could only think differently.

Advocates, nonprofit directors, waste managers, landfill directors, environmentalists, entrepreneurs, eco-farmers and anyone searching for environmental solutions and opportunities with source separation should not only read this book ... we must figure out how to change our thinking, and put Gillespie's plan into action.

#7582 • Softcover • Copyright 2020 • 177 pages • $28.00

www.ingramcontent.com/pod-product-compliance
Lightning Source LLC
Chambersburg PA
CBHW030820270326
41928CB00007B/824